SpringerBriefs in Mathematics

Series Editors

Nicola Bellomo
Michele Benzi
Palle E.T. Jorgensen
Tatsien Li
Roderick Melnik
Otmar Scherzer
Benjamin Steinberg
Lothar Reichel
Yuri Tschinkel
G. George Yin
Ping Zhang

SpringerBriefs in Mathematics showcases expositions in all areas of mathematics and applied mathematics. Manuscripts presenting new results or a single new result in a classical field, new field, or an emerging topic, applications, or bridges between new results and already published works, are encouraged. The series is intended for mathematicians and applied mathematicians.

More information about this series at http://www.springer.com/series/10030

Bernard Bercu • Bernard Delyon • Emmanuel Rio

Concentration Inequalities for Sums and Martingales

 Springer

Bernard Bercu
Institut de Mathématiques
  de Bordeaux
Université de Bordeaux
Talence, France

Emmanuel Rio
Laboratoire de Mathématiques
  de Versailles
Université de Versailles
  St. Quentin en Yvelines
Versailles, France

Bernard Delyon
Institut de Recherche Mathématique
  de Rennes
Université de Rennes
Rennes, France

ISSN 2191-8198   ISSN 2191-8201   (electronic)
SpringerBriefs in Mathematics
ISBN 978-3-319-22098-7   ISBN 978-3-319-22099-4   (eBook)
DOI 10.1007/978-3-319-22099-4

Library of Congress Control Number: 2015945946

Mathematics Subject Classification (2010): 60-01, 60E15, 60F10, 60G42, 60G50

Springer Cham Heidelberg New York Dordrecht London

Printed on acid-free paper

Springer International Publishing AG Switzerland is part of Springer Science+Business Media (www.springer.com)

In memory of our friend Abderrahmen Touati

Preface

Over the last two decades, there has been a renewed interest in the area of concentration inequalities. The starting point of this short book was a project on exponential inequalities for martingales with a view toward applications in probability and statistics. During the preparation of this book, we realized that the classical exponential inequalities for sums of independent random variables were not well reported in the literature. This motivated us to write a chapter entirely devoted to sums of independent random variables, which includes the classical deviation inequalities of Bernstein, Bennett, and Hoeffding as well as less-recognized inequalities and new results. Some of these inequalities are extended to martingales in the third chapter, which deals with concentration inequalities for martingales and self-normalized martingales. We end this book with a brief chapter devoted to a few applications in probability and statistics, which shows the striking efficiency of martingales techniques on some examples. We wish to emphasize that this short book does not provide a complete overview of martingale exponential inequalities and their applications. More sophisticated results can be found in the literature. We hope that researchers interested in concentration inequalities for sums and martingales will find in this book useful tools for their future research.

Talence, France
Rennes, France
Versailles, France
June 2015

Bernard Bercu
Bernard Delyon
Emmanuel Rio

Contents

Chapter 1
Classical results

1.1 Sums of independent random variables

1.1.1 Strong law of large numbers for sums

Let (X_n) be a sequence of independent and identically distributed integrable random variables with common mean m. Assume that m is unknown. An exact confidence interval for m can be calculated via the sample mean

$$\overline{X}_n = \frac{1}{n} S_n \qquad \text{where} \qquad S_n = \sum_{k=1}^{n} X_k.$$

First of all, the strong law of large numbers ensures that m can be properly estimated by the sample mean.

Theorem 1.1 (Strong law of large numbers for sums). *Let (X_n) be a sequence of independent and identically distributed integrable random variables with mean m. Then, we have*

$$\lim_{n \to \infty} \overline{X}_n = m \qquad a.s. \tag{1.1}$$

A first evaluation of the mean m can be obtained from the strong law of large numbers. However, it is impossible to deduce a confidence interval for m from Theorem 1.1. An asymptotic confidence interval can be constructed from the central limit theorem.

© The Authors 2015

B. Bercu et al., *Concentration Inequalities for Sums and Martingales*,
SpringerBriefs in Mathematics, DOI 10.1007/978-3-319-22099-4_1

1.1.2 Central limit theorem for sums

The central limit theorem for sums of independent and identically distributed random variables is as follows.

Theorem 1.2 (Central limit theorem for sums). *Let (X_n) be a sequence of independent and identically distributed square integrable random variables with mean m and positive variance σ^2. Then, we have*

$$\sqrt{n}(\overline{X}_n - m) \overset{\mathcal{L}}{\longrightarrow} \mathcal{N}(0, \sigma^2). \tag{1.2}$$

In most of all situations, the variance σ^2 is unknown. It is necessary to estimate it by the sample variance

$$\hat{\sigma}_n^2 = \frac{1}{n} \sum_{k=1}^{n} (X_k - \overline{X}_n)^2.$$

As $\hat{\sigma}_n^2$ converges almost surely to σ^2, one can deduce from Theorem 1.2 that

$$\sqrt{n} \left(\frac{\overline{X}_n - m}{\hat{\sigma}_n} \right) \overset{\mathcal{L}}{\longrightarrow} \mathcal{N}(0, 1).$$

Therefore, an asymptotic confidence interval for m is given by

$$\mathscr{I}(m) = \left[\overline{X}_n - a\frac{\hat{\sigma}_n}{\sqrt{n}}, \ \overline{X}_n + a\frac{\hat{\sigma}_n}{\sqrt{n}} \right]$$

where, for a confidence level $1 - \alpha$ with $0 < \alpha < 1$, the value a stands for the $1 - \alpha/2$ quantile of the $\mathcal{N}(0, 1)$ distribution. Another strategy is to make use of the large deviation properties for the sample mean. This is doomed to fail because the rate function in the large deviation principle for the sample mean mostly depends on the mean m as well as on the variance σ^2.

1.1.3 Large deviations

The large deviation properties for the sample mean are as follows.

Theorem 1.3 (Cramér-Chernoff's Theorem). *Let (X_n) be a sequence of independent and identically distributed random variables. Denote by I the Legendre-Fenchel transform of the log-Laplace ℓ of (X_n), namely*

$$I(x) = \sup_{t \in \mathbb{R}} \{ xt - \ell(t) \}.$$

Then, the sequence $(\overline{X}_n)$ satisfies a large deviation principle with good rate function I which means that, for any closed set $F \subset \mathbb{R}$,

$$\limsup_{n\to\infty} \frac{1}{n} \log \mathbb{P}\left(\overline{X}_n \in F\right) \leqslant - \inf_{x \in F} I(x), \tag{1.3}$$

while, for any open set $G \subset \mathbb{R}$,

$$\liminf_{n\to\infty} \frac{1}{n} \log \mathbb{P}\left(\overline{X}_n \in G\right) \geqslant - \inf_{x \in G} I(x). \tag{1.4}$$

Remark 1.4. One can observe that Theorem 1.3 holds even when the sequence (X_n) is not integrable. However, in this case, the rate function I is identically zero. Moreover, if the sequence (X_n) is integrable with mean m, then $I(m) = 0$ and for all $x > m$,

$$\lim_{n\to\infty} \frac{1}{n} \log \mathbb{P}\left(\overline{X}_n \geqslant x\right) = -I(x),$$

whereas, for all $x < m$,

$$\lim_{n\to\infty} \frac{1}{n} \log \mathbb{P}\left(\overline{X}_n \leqslant x\right) = -I(x).$$

Finally, the usual Chernoff calculation reveals that, for all $n \geqslant 1$ and for all $x > m$,

$$\mathbb{P}\left(\overline{X}_n \geqslant x\right) \leqslant \exp(-nI(x)),$$

while, for all $n \geqslant 1$ and for all $x < m$,

$$\mathbb{P}\left(\overline{X}_n \leqslant x\right) \leqslant \exp(-nI(x)).$$

One can see below that in most of all cases, the rate function I depends on the mean m. The reader is referred to Dembo-Zeitouni [3] for an exposition of the theory of large deviations with applications.

Example 1.5 (Discrete random variables).

1) If X has a Bernoulli $\mathscr{B}(p)$ distribution with $0 < p < 1$, then for all $x \in [0, 1]$

$$I(x) = x \log\left(\frac{x}{p}\right) + (1 - x) \log\left(\frac{1 - x}{1 - p}\right)$$

and $I(x) = +\infty$ otherwise.

2) If Y has a Poisson $\mathscr{P}(\lambda)$ distribution with $\lambda > 0$, then for all $x \geqslant 0$

$$I(x) = x \log\left(\frac{x}{\lambda}\right) - x + \lambda$$

and $I(x) = +\infty$ otherwise.

Example 1.6 (Continuous random variables).

1) If X has an Exponential $\mathscr{E}(\lambda)$ distribution with $\lambda > 0$, then for all $x > 0$

$$I(x) = \lambda x - 1 - \log(\lambda x)$$

and $I(x) = +\infty$ otherwise.

2) If X has a Normal $\mathscr{N}(m, \sigma^2)$ distribution with $m \in \mathbb{R}$ and $\sigma^2 > 0$, then for all $x \in \mathbb{R}$

$$I(x) = \frac{(x - m)^2}{2\sigma^2}$$

Actually, it is much more efficient in applications to make use of sharp large deviations, see Bahadur-Rao [1].

Theorem 1.7 (Bahadur-Rao's Theorem). *Let (X_n) be a sequence of independent and identically distributed random variables. Assume that the log-Laplace ℓ of (X_n) is finite on all $\mathbb{R}$ and that the distribution of (X_n) is absolutely continuous. Then, the sequence $(\overline{X}_n)$ satisfies a sharp large deviation principle. In particular, for all $x \in \mathbb{R}$, it exists a sequence $(d_k(x))$ such that for any $p \geqslant 1$ and n large enough, if $x > m$*

$$\mathbb{P}(\overline{X}_n \geqslant x) = \frac{\exp(-nI(x))}{\sigma_x t_x \sqrt{2\pi n}} \left[1 + \sum_{k=1}^{p} \frac{d_k(x)}{n^k} + \mathcal{O}\left(\frac{1}{n^{p+1}}\right) \right], \tag{1.5}$$

whereas, if $x < m$

$$\mathbb{P}(\overline{X}_n \leqslant x) = -\frac{\exp(-nI(x))}{\sigma_x t_x \sqrt{2\pi n}} \left[1 + \sum_{k=1}^{p} \frac{d_k(x)}{n^k} + \mathcal{O}\left(\frac{1}{n^{p+1}}\right) \right], \tag{1.6}$$

where the value t_x is given by $\ell'(t_x) = x$ and $\sigma_x^2 = \ell''(t_x)$. Finally, all the coefficients $d_k(x)$ may be explicitly calculated as functions of the derivatives of the function ℓ at point t_x.

One can realize that the strategy to compute exact confidence interval for m via the sample mean is far from being obvious and understood. We shall now restrict ourselves to the case of bounded random variables, which is much more easy to handle. For the sake of simplicity, assume that (X_n) be a sequence of independent and identically distributed random variables with Bernoulli $\mathscr{B}(p)$ distribution with $0 < p < 1$. We clearly have $m = p$ and $\sigma^2 = p(1 - p)$. In order to calculate a confidence interval for p, a first naive approach is related to Markov's inequality given, for all $a > 0$, by

$$\mathbb{P}(|\overline{X}_n - p| \geqslant a) \leqslant \frac{p(1 - p)}{na^2}.$$

As $4p(1 - p) \leqslant 1$, we clearly have

$$\mathbb{P}(|\overline{X}_n - p| \leqslant a) \geqslant 1 - \frac{1}{4na^2}.$$

However, as soon as $n > 1/4a^2$, $0 < \alpha < 1$ where $\alpha = 1/4na^2$. Hence, a first exact confidence interval for p, with confidence level $1 - \alpha$, is given by

$$\mathscr{I}(p) = \left[\overline{X}_n - \frac{1}{2\sqrt{n\alpha}}, \ \overline{X}_n + \frac{1}{2\sqrt{n\alpha}} \right].$$

An alternative approach is to make use of the central limit theorem. We have

$$\overline{X}_n - p = \frac{\sqrt{p(1-p)}}{\sqrt{n}} Y_n \qquad \text{where} \qquad Y_n = \frac{\sqrt{n}(\overline{X}_n - p)}{\sqrt{p(1-p)}}.$$

We deduce from Theorem 1.2 that Y_n converges in distribution to an $\mathscr{N}(0,1)$ random variable. Moreover, for all $a > 0$, we have $\mathbb{P}(|\overline{X}_n - p| \leqslant a) \geqslant \mathbb{P}(|Y_n| \leqslant 2a\sqrt{n})$. Consequently, a second asymptotic confidence interval for p, with confidence level $1 - \alpha$, is given by

$$\mathscr{J}(p) = \left[\overline{X}_n - \frac{a}{2\sqrt{n}}, \ \overline{X}_n + \frac{a}{2\sqrt{n}} \right]$$

where the value a stands for the $1 - \alpha/2$ quantile of the $\mathscr{N}(0,1)$ distribution. A third approach relies on the almost sure convergence

$$\lim_{n \to \infty} \overline{X}_n (1 - \overline{X}_n) = p(1-p) \qquad \text{a.s.}$$

Slutsky's theorem implies that

$$\sqrt{n} \frac{\overline{X}_n - p}{\sqrt{\overline{X}_n(1 - \overline{X}_n)}} \overset{\mathcal{L}}{\longrightarrow} \mathscr{N}(0,1).$$

Therefore, a third asymptotic confidence interval for p is given by

$$\mathscr{K}(p) = \left[\overline{X}_n - a\sqrt{\frac{\overline{X}_n(1 - \overline{X}_n)}{n}}, \ \overline{X}_n + a\sqrt{\frac{\overline{X}_n(1 - \overline{X}_n)}{n}} \right]$$

where a is the $1 - \alpha/2$ quantile of the $\mathscr{N}(0,1)$ distribution. Finally, we shall see from Hoeffding's inequality for sums of independent and bounded random variables that for all $a > 0$,

$$\mathbb{P}(|\overline{X}_n - p| \geqslant a) \leqslant 2\exp(-2na^2).$$

It ensures that, for $n > (2a^2)^{-1}\log 2$,

$$\mathbb{P}(|\overline{X}_n - p| \leqslant a) \geqslant 1 - 2\exp(-2na^2) = 1 - \alpha$$

where $\alpha = 2\exp(-2na^2)$ lies in $]0,1[$. Consequently, a fourth exact confidence interval for p, with confidence level $1-\alpha$, is given by

$$\mathscr{H}(p) = \left[\overline{X}_n - \sqrt{\frac{\log(2/\alpha)}{2n}},\ \overline{X}_n + \sqrt{\frac{\log(2/\alpha)}{2n}}\right].$$

The four confidence intervals for p are illustrated in Figure 1.1 with the standard choice $\alpha = 5\%$ and for n varying from 10 to 100. On the one side, the confidence intervals $\mathscr{J}(p)$ and $\mathscr{K}(p)$ are always more accurate than $\mathscr{I}(p)$ and $\mathscr{H}(p)$, providing of course that the Gaussian approximation is justified. On the other side, the bounds given by $\mathscr{I}(p)$ and $\mathscr{H}(p)$ are always true whatever is the value of n. One can easily understand in this example the practical interest of concentration inequalities in Statistical applications.

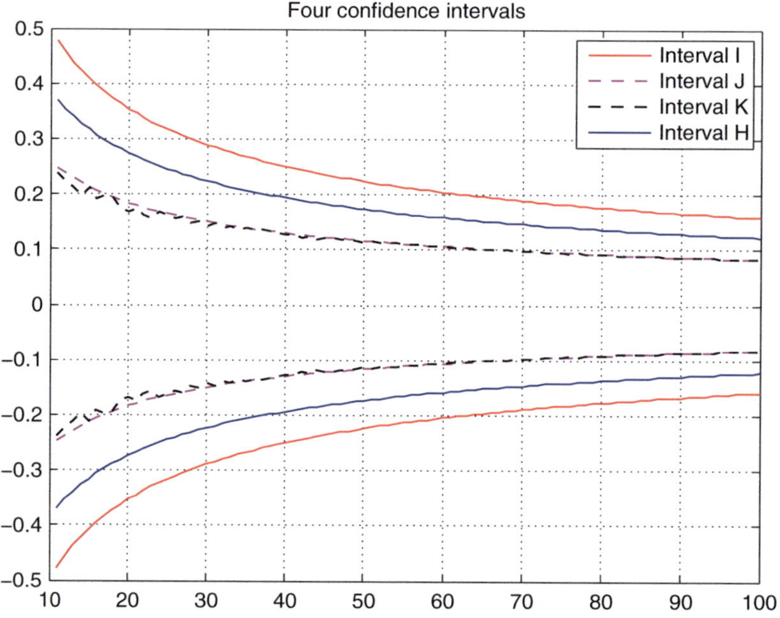

Fig. 1.1 Comparisons between confidence intervals

1.2 Martingales

We shall now focus our attention on asymptotic results for martingales. Let $(\Omega, \mathscr{A}, \mathbb{P})$ be a probability space endowed with a filtration $\mathbb{F} = (\mathscr{F}_n)$ which means that $\mathbb{F}$ is an increasing sequence of sub-σ-algebras $\mathscr{F}_n$ of $\mathscr{A}$. One can see $\mathscr{F}_n$ as the σ-algebra of events occurring up to time n.

Definition 1.8. Let (M_n) be a sequence of integrable random variables defined on $(\Omega, \mathscr{A}, \mathbb{P})$ such that, for all $n \geqslant 0$, M_n is $\mathscr{F}_n$-measurable. We shall say that (M_n) is a martingale, submartingale, or supermartingale if, for all $n \geqslant 0$, we have almost surely

$$\mathbb{E}[M_{n+1} | \mathscr{F}_n] = M_n, \qquad \mathbb{E}[M_{n+1} | \mathscr{F}_n] \geqslant M_n, \qquad \mathbb{E}[M_{n+1} | \mathscr{F}_n] \leqslant M_n.$$

Two examples related to sum and product martingales are as follows.

Example 1.9 (Sum martingale). Let (X_n) be a sequence of integrable and independent random variables such that, for all $n \geqslant 1$, $\mathbb{E}[X_n] = m$. Denote

$$S_n = \sum_{k=1}^{n} X_k.$$

Then, (S_n) is a martingale if $m = 0$, submartingale if $m \geqslant 0$, supermartingale if $m \leqslant 0$.

Example 1.10 (Product martingale). Let (X_n) be a sequence of positive, integrable, and independent random variables such that, for all $n \geqslant 1$, $\mathbb{E}[X_n] = m$. Denote

$$P_n = \prod_{k=1}^{n} X_k.$$

Then, (P_n) is a martingale if $m = 1$, submartingale if $m \geqslant 1$, supermartingale if $m \leqslant 1$.

It is well known in Analysis that every increasing sequence bounded from above converges to its supremum, while every decreasing sequence bounded from below converges to its infimum. The stochastic analogous of this result is due to Doob.

Lemma 1.11 (Doob's Lemma). *If (M_n) is a submartingale bounded from above by some constant M, then (M_n) converges almost surely. It is also the case if (M_n) is a supermartingale bounded from below by some constant m.*

The famous Doob's convergence theorem for martingales is as follows.

Theorem 1.12 (Doob's convergence theorem). *Let (M_n) be a martingale, submartingale, or supermartingale, bounded in $\mathbb{L}^1$ which means*

$$\sup_{n \geqslant 0} \mathbb{E}[|M_n|] < +\infty.$$

Then, (M_n) converges a.s. to an integrable random variable.

Theorem 1.13. *Let (M_n) be a martingale bounded in $\mathbb{L}^p$ with $p \geqslant 1$, which means that*

$$\sup_{n \geqslant 0} \mathbb{E}[|M_n|^p] < +\infty.$$

1) If $p > 1$, then (M_n) converges a.s. to an integrable random variable. The convergence is also true in $\mathbb{L}^p$.

2) If $p = 1$, (M_n) converges a.s. to an integrable random variable. The convergence holds in $\mathbb{L}^1$ as soon as (M_n) is uniformly integrable, that is

$$\lim_{a \to \infty} \sup_{n \geq 0} \mathbb{E}[|M_n| \mathbb{I}_{\{|M_n| \geq a\}}] = 0.$$

Example 1.14 (Exponential martingale). Let (X_n) be a sequence of independent random variables sharing the same $\mathscr{N}(0,1)$ distribution. Let $S_n = X_1 + \cdots + X_n$ and denote for any real t different from zero,

$$M_n(t) = \exp\left(tS_n - \frac{nt^2}{2}\right).$$

It is clear that $(M_n(t))$ is a martingale which converges almost surely to zero as S_n/n goes to zero almost surely. However, $\mathbb{E}[M_n(t)] = \mathbb{E}[M_1(t)] = 1$ which means that $(M_n(t))$ does not converge in $\mathbb{L}^1$.

Example 1.15 (Autoregressive martingale). Let (X_n) be the autoregressive process given, for all $n \geq 0$, by

$$X_{n+1} = \theta X_n + (1 - \theta)\varepsilon_{n+1}$$

where $X_0 = p$ with $0 < p < 1$ and the parameter $0 < \theta < 1$. Assume that the conditional distribution $\mathscr{L}(\varepsilon_{n+1}|\mathscr{F}_n)$ is the Bernoulli $\mathscr{B}(X_n)$ distribution. It is not hard to see by induction that $0 < X_n < 1$. Moreover, (X_n) is a martingale bounded in $\mathbb{L}^1$ which converges almost surely to a random variable X. This convergence also holds in $\mathbb{L}^1$ and X has the Bernoulli $\mathscr{B}(p)$ distribution.

1.2.1 Strong law of large numbers for martingales

We are now interested in the strong law of large numbers for square integrable martingales. We refer the reader to [4–7] for more insight on the theory of discrete-time martingales with many applications in Probability and Statistics.

Definition 1.16. Let (M_n) be a square integrable martingale which means that for all $n \geq 1$, $\mathbb{E}[M_n^2] < \infty$. The increasing process associated with (M_n) is defined by $\langle M \rangle_0 = 0$ and, for all $n \geq 1$,

$$\langle M \rangle_n = \sum_{k=1}^{n} \mathbb{E}[(M_k - M_{k-1})^2 | \mathscr{F}_{k-1}].$$

We first deal with a local convergence theorem for martingales due to Chow [2]. It allows to avoid the assumption that the martingale is square integrable.

Theorem 1.17 (Chow's convergence theorem). *Let (M_n) be a martingale such that for some $1 \leqslant p \leqslant 2$ and for all $n \geqslant 1$, $\mathbb{E}[|M_n|^p] < \infty$. On the set*

$$\Gamma = \left\{ \sum_{n=0}^{\infty} \mathbb{E}[|\Delta M_n|^p | \mathscr{F}_{n-1}] < \infty \right\}$$

where $\Delta M_n = M_n - M_{n-1}$, (M_n) converges almost surely.

The strong law of large numbers for martingales is as follows.

Theorem 1.18 (Strong law of large numbers for martingales). *Let (M_n) be a square integrable martingale and denote by $(\langle M \rangle_n)$ its increasing process. Let*

$$\langle M \rangle_\infty = \lim_{n \to \infty} \langle M \rangle_n.$$

1) On the set $\left\{ \langle M \rangle_\infty < \infty \right\}$, (M_n) converges almost surely to a square integrable random variable.
2) On the set $\left\{ \langle M \rangle_\infty = \infty \right\}$, we have

$$\lim_{n \to \infty} \frac{M_n}{\langle M \rangle_n} - 0 \qquad a.s. \tag{1.7}$$

More precisely, for any positive γ,

$$\frac{M_n^2}{\langle M \rangle_n} = o\left(\left(\log \langle M \rangle_n \right)^{1+\gamma} \right) \qquad a.s. \tag{1.8}$$

Remark 1.19. If it exists a sequence (a_n) of positive real numbers, increasing to infinity, such that $\langle M \rangle_n = O(a_n)$, then we have $M_n = o(a_n)$ a.s.

1.2.2 Central limit theorem for martingales

We achieve this introductory chapter by the central limit theorem for martingales.

Theorem 1.20 (Central limit theorem for martingales). *Let (M_n) be a square integrable martingale and let (a_n) be a sequence of positive real numbers increasing to infinity. Assume that*

1) It exists a deterministic limit $\ell \geqslant 0$ such that

$$\frac{\langle M \rangle_n}{a_n} \xrightarrow{\mathcal{P}} \ell.$$

2) Lindeberg's condition is satisfied which means that for all $\varepsilon > 0$,

$$\frac{1}{a_n} \sum_{k=1}^{n} \mathbb{E}[|\Delta M_k|^2 I_{\{|\Delta M_k| \geqslant \varepsilon \sqrt{a_n}\}} \,|\, \mathscr{F}_{k-1}] \xrightarrow{\mathcal{P}} 0.$$

Then, we have

$$\frac{1}{\sqrt{a_n}} M_n \xrightarrow{\mathcal{L}} \mathcal{N}(0, \ell). \tag{1.9}$$

Moreover, if $\ell > 0$, we also have

$$\sqrt{a_n}\left(\frac{M_n}{\langle M \rangle_n}\right) \xrightarrow{\mathcal{L}} \mathcal{N}(0, \ell^{-1}). \tag{1.10}$$

References

1. Bahadur, R. R. and Ranga Rao, R.: On deviations of the sample mean. Ann. Math. Statist. **31**, 1015–1027 (1960)
2. Chow, Y. S.: Local convergence of martingales and the law of large numbers. Ann. Math. Statist. **36**, 552–558 (1965)
3. Dembo, A. and Zeitouni, O.: Large deviations techniques and applications. Springer-Verlag, New York, second edition (1998)
4. Duflo, M.: Random iterative models. Springer-Verlag, Berlin (1997)
5. Hall, P. and Heyde, C. C.: Martingale limit theory and its application. Academic Press Inc., New York (1980)
6. Kushner, H. J. and Yin, G. G.: Stochastic approximation and recursive algorithms and applications. Springer-Verlag, New York (2003)
7. Neveu, J.: Discrete-parameter martingales. North-Holland Publishing Co., New York (1975)

Chapter 2
Concentration inequalities for sums

2.1 Bernstein's inequalities

Sergei Bernstein [6] was at the beginning of exponential inequalities for sums of independent random variables. This section deals with Bernstein's type inequalities. It is divided into three subsections. In the first one, we focus our attention on the one-sided inequality of Bernstein [6] as well as on improvements of this inequality. The second one is devoted to two-sided versions of Bernstein's inequality. In the last one, we give new inequalities with a smaller second term.

2.1.1 One-sided inequalities

The main result of this subsection is the theorem below, which collects different versions and improvements of Bernstein's one-sided inequality.

Theorem 2.1. *Let $X_1, \ldots, X_n$ be a finite sequence of independent random variables with finite variances. Denote*

$$S_n = X_1 + \cdots + X_n, \qquad \mathscr{V}_n = \mathbb{E}[X_1^2] + \cdots + \mathbb{E}[X_n^2],$$

$$v_n = \frac{\mathscr{V}_n}{n}. \tag{2.1}$$

Assume that $\mathbb{E}[S_n] = 0$ and that there exists some positive constant c such that, for any integer $p \geqslant 3$,

$$\sum_{k=1}^{n} \mathbb{E}\left[(\max(0, X_k))^p\right] \leqslant \frac{p! c^{p-2}}{2} \mathscr{V}_n. \tag{2.2}$$

© The Authors 2015

B. Bercu et al., *Concentration Inequalities for Sums and Martingales*,

SpringerBriefs in Mathematics, DOI 10.1007/978-3-319-22099-4_2

Then, for any positive x,

$$\mathbb{P}(S_n \geqslant nx) \leqslant \left(1 + \frac{x^2}{2(v_n + cx)}\right)^n \exp\left(-\frac{nx^2}{v_n + cx}\right) \tag{2.3}$$

$$\leqslant \exp\left(-\frac{nx^2}{2(v_n + cx)}\right). \tag{2.4}$$

In addition, we also have, for any positive x,

$$\mathbb{P}(S_n \geqslant nx) \leqslant \exp\left(-\frac{nx^2}{v_n + cx + \sqrt{v_n(v_n + 2cx)}}\right) \tag{2.5}$$

and

$$\mathbb{P}\big(S_n > n(cx + \sqrt{2v_n x})\big) \leqslant \exp(-nx). \tag{2.6}$$

Remark 2.2. It is not necessary to assume that the random variables $X_1, \ldots, X_n$ are centered. We only have to suppose that $\mathbb{E}[S_n] = 0$. In the centered case, $\mathcal{V}_n$ coincides with $V_n = \mathrm{Var}(S_n)$. Otherwise, $\mathcal{V}_n$ is obviously larger than V_n.

Remark 2.3. Condition (2.2), given in Rio [22], is weaker than the standard Bernstein's condition, which says that for any $1 \leqslant k \leqslant n$ and for any integer $p \geqslant 3$,

$$\mathbb{E}\big[|X_k|^p\big] \leqslant \frac{p!c^{p-2}}{2} \mathbb{E}[X_k^2].$$

For example, condition (2.2) allows to consider random variables with heavier tails on the left. Bernstein [6] proved (2.4) under the above condition.

Remark 2.4. The optimal constant c^* in Theorem 2.1 is the smallest positive real c such that condition (2.2) is satisfied. If the random variables $X_1, \ldots, X_n$ are such that, for all $1 \leqslant k \leqslant n$, $X_k \leqslant b$ almost surely for some positive constant b, then one can prove that $c^* \leqslant b/3$. In the forthcoming sections, we will give more efficient inequalities for random variables bounded from above.

Remark 2.5. Inequality (2.5) was obtained by Bennett [2] and it was called first improvement of Bernstein's inequality. Nevertheless, this result is still suboptimal. One can observe that (2.5) and (2.6) are equivalent. In the case $v_n \geqslant c^2$, inequality (2.5) is less efficient than (2.3) as shown in Figure 2.1 as well as in Figure 2.2 which compare the rate functions

$$\varphi(x) = \frac{x^2}{v_n + cx} - \log\left(1 + \frac{x^2}{2(v_n + cx)}\right)$$

$$\Phi(x) = \frac{x^2}{v_n + cx + \sqrt{v_n(v_n + 2cx)}} \quad \text{and} \quad \Psi(x) = \frac{x^2}{2(v_n + cx)}$$

associated, respectively, to the new Bernstein's inequality, Bennett's and Bernstein's inequalities, in the particular cases $v_n = 1$, $c = 1$, and $v_n = 1$, $c = 1/2$, respectively. Note also that (2.4) is equivalent to the reverse inequality

$$\mathbb{P}\big(S_n \geqslant n(cx + \sqrt{2v_n x + (cx)^2})\big) \leqslant \exp(-nx),$$

which is less efficient than (2.6).

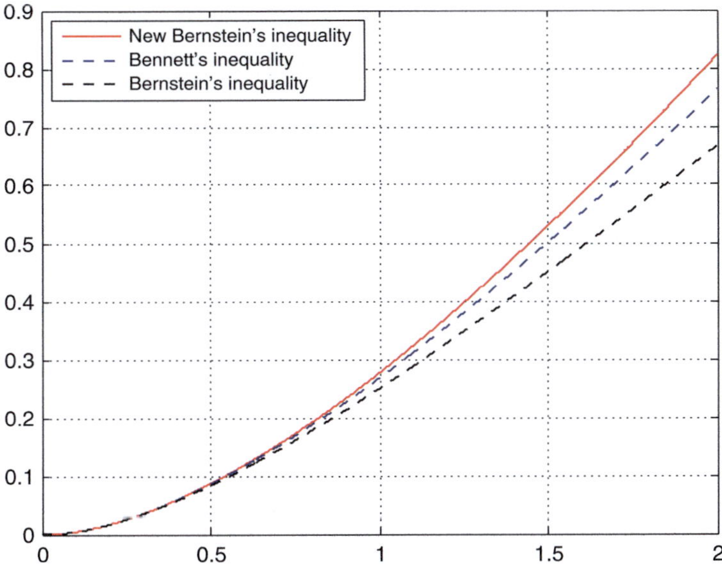

Fig. 2.1 Comparisons in Bernstein's inequalities in the particular case $v_n = 1$ and $c = 1$.

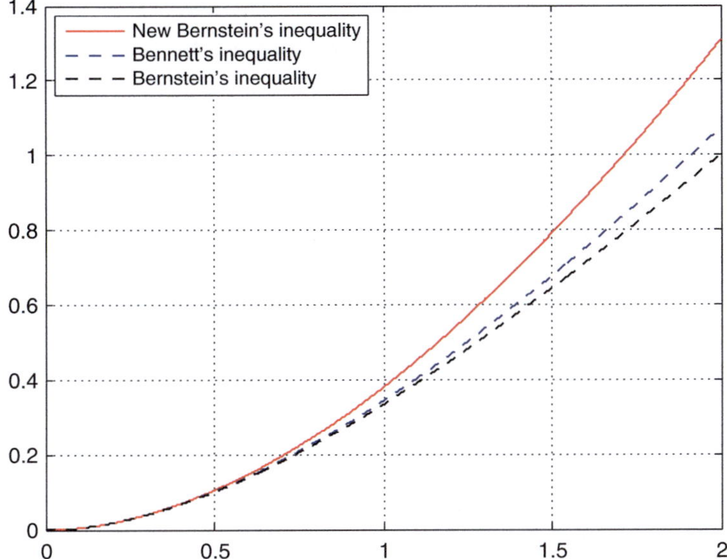

Fig. 2.2 Comparisons in Bernstein's inequalities in the particular case $v_n = 1$ and $c = 1/2$.

The outline of the proof of Theorem 2.1 is as follows. Denote by ℓ an upper bound of the normalized cumulant generating function of S_n, that is, for all $t \geqslant 0$,

$$\frac{1}{n} \log \mathbb{E}[\exp(tS_n)] \leqslant \ell(t).$$

We immediately obtain from Markov's inequality that, for all $t \geqslant 0$ and $x \geqslant 0$,

$$\mathbb{P}(S_n \geqslant nx) \leqslant \exp(-n(tx - \ell(t))), \tag{2.7}$$

which leads to

$$\mathbb{P}(S_n \geqslant nx) \leqslant \exp(-n\ell^*(x)) \quad \text{where} \quad \ell^*(x) = \sup_{t \geqslant 0}(xt - \ell(t)). \tag{2.8}$$

Consequently, in order to establish a concentration inequality for S_n, it only remains to calculate the Legendre-Fenchel transform ℓ^* of ℓ. In many situations, the calculation of ℓ^* could be somehow complicated. However, a sharp upper bound for ℓ^* also leads to concentration inequalities. Another strategy is to realize that for any positive x,

$$\mathbb{P}\left(S_n > n \inf_{t>0}\left(\frac{\ell(t)+x}{t}\right)\right) = \sup_{t>0}\mathbb{P}\left(S_n > n\left(\frac{\ell(t)+x}{t}\right)\right) \leqslant \exp(-nx), \tag{2.9}$$

which gives an upper bound for the quantile function of S_n in a more direct way. The proof of Theorem 2.1 relies on two elementary lemmas. The first one, due to Hoeffding [11], allows us to replace the initial random variables by independent random variables with the same distribution.

Lemma 2.6. *Let $X_1, \ldots X_n$ be a finite sequence of independent random variables and denote $S_n = X_1 + \cdots + X_n$. Then, for any real t,*

$$\frac{1}{n} \log \mathbb{E}[\exp(tS_n)] \leqslant \ell(t) \quad \text{where} \quad \ell(t) = \log\left(\frac{1}{n} \sum_{k=1}^{n} \mathbb{E}[\exp(tX_k)]\right). \tag{2.10}$$

Proof. It follows from the independence of the random variables $X_1, \ldots X_n$ that for any real t,

$$\log \mathbb{E}[\exp(tS_n)] = \sum_{k=1}^{n} \log \mathbb{E}[\exp(tX_k)].$$

The concavity of the logarithm function clearly leads to

$$\frac{1}{n} \log \mathbb{E}[\exp(tS_n)] \leqslant \log\left(\frac{1}{n} \sum_{k=1}^{n} \mathbb{E}[\exp(tX_k)]\right) = \ell(t),$$

which achieves the proof of Lemma 2.6. $\qquad\qquad\square$

The second lemma, which implies (2.9), will allow us to give a nice proof of the first improvement of Bennett.

Lemma 2.7. *Let X be a real valued random variable with a finite Laplace transform on a right neighborhood of the origin. Denote by L_X the logarithm of the Laplace transform of X. Then, for any positive real number x,*

$$\mathbb{P}\Big(X > \inf_{t>0}\Big(\frac{L_X(t)+x}{t}\Big)\Big) \leqslant \exp(-x). \tag{2.11}$$

Proof. Let t be any positive real number such that $L_X(t) < \infty$. We immediately deduce from Markov's inequality that for any positive real number x,

$$\mathbb{P}\Big(X \geqslant \Big(\frac{L_X(t)+x}{t}\Big)\Big) = \mathbb{P}\Big(\exp(tX) \geqslant \exp(L_X(t)+x)\Big) \leqslant \exp(-x). \tag{2.12}$$

Hence, we obtain (2.11) by taking the infimum over all positive real numbers t. $\square$

Proof of Theorem 2.1. We are now in position to prove Theorem 2.1. First of all, one can observe that for any $x \leqslant 0$,

$$\exp(x) \leqslant 1 + x + \frac{x^2}{2},$$

which ensures that for any real number x,

$$\exp(x) \leqslant 1 + x + \frac{x^2}{2} + \sum_{p=3}^{\infty} \frac{(x_+)^p}{p!}$$

where $x_+ = \max(0,x)$ stands for the positive part of x. It follows from the monotone convergence theorem that for all $1 \leqslant k \leqslant n$ and for any positive t,

$$\mathbb{E}[\exp(tX_k)] \leqslant 1 + t\mathbb{E}[X_k] + \frac{t^2\mathbb{E}[X_k^2]}{2} + \sum_{p=3}^{\infty} \frac{t^p\mathbb{E}[(\max(0,X_k))^p]}{p!}. \tag{2.13}$$

Consequently, as $\mathbb{E}[S_n] = 0$, we deduce from Bernstein's condition (2.2) that

$$\sum_{k=1}^{n} \mathbb{E}[\exp(tX_k)] \leqslant n + \frac{v_n}{2} \sum_{p=2}^{\infty} c^{p-2}t^p.$$

Hence, as soon as $0 < tc < 1$,

$$\exp(\ell(t)) = \frac{1}{n}\sum_{k=1}^{n} \mathbb{E}[\exp(tX_k)] \leqslant 1 + \frac{v_n t^2}{2(1-tc)}.$$

Therefore, we find from Lemma 2.6 that for any positive t such that $0 < tc < 1$,

$$\frac{1}{n}\log\mathbb{E}[\exp(tS_n)] \leqslant \ell_{v_n}(t) \quad \text{where} \quad \ell_{v_n}(t) = \log\left(1 + \frac{v_n t^2}{2(1 - tc)}\right). \tag{2.14}$$

Hereafter, if we choose $t_n(x) = x/(v_n + cx)$ with $x > 0$, we obtain from the definition of the Legendre-Fenchel transform that

$$\ell_{v_n}^*(x) \geqslant x t_n(x) - \ell_{v_n}(t_n(x)) = \frac{x^2}{v_n + cx} - \log\left(1 + \frac{v_n t_n^2(x)}{2(1 - t_n(x)c)}\right),$$

$$= \frac{x^2}{v_n + cx} - \log\left(1 + \frac{x^2}{2(v_n + cx)}\right). \tag{2.15}$$

Then, (2.3) and (2.4) immediately follow from (2.8) and (2.15). Finally, it only remains to prove (2.6). Applying Lemma 2.7 to S_n, we obtain that

$$\mathbb{P}\left(S_n > n \inf_{tc \in]0,1[}\left(\frac{\ell_{v_n}(t) + x}{t}\right)\right) \leqslant \exp(-nx). \tag{2.16}$$

Now, it follows from (2.14) and the elementary inequality $\log(1 + x) \leqslant x$ that

$$\inf_{tc \in]0,1[}\left(\frac{\ell_{v_n}(t) + x}{t}\right) \leqslant \inf_{tc \in]0,1[}\left(\frac{v_n t}{2(1 - tc)} + \frac{x}{t}\right) = cx + \sqrt{2xv_n} \tag{2.17}$$

where the infimum of the right-hand side is given by the optimal value

$$t = \frac{\sqrt{2x}}{\sqrt{v_n} + c\sqrt{2x}}.$$

Finally, we clearly deduce (2.6) from (2.16) and (2.17), which completes the proof of Theorem 2.1. $\qquad\square$

Example 2.8. Let $\varepsilon_1, \ldots, \varepsilon_n$ be a finite sequence of independent random variables sharing the same Exponential $\mathscr{E}(\lambda)$ distribution with $\lambda > 0$. Let $a = (a_1, \ldots, a_n)$ be a vector with positive real components. For all $1 \leqslant k \leqslant n$, denote

$$X_k = a_k\left(\varepsilon_k - \frac{1}{\lambda}\right).$$

One can easily check that

$$\mathscr{V}_n = \sum_{k=1}^{n}\mathbb{E}[X_k^2] = \frac{1}{\lambda^2}\sum_{k=1}^{n}a_k^2 = \frac{\|a\|_2^2}{\lambda^2}.$$

In addition, it is not hard to see that condition (2.2) holds true with

$$c = \frac{1}{\lambda}\max(a_1, \ldots, a_n) = \frac{\|a\|_\infty}{\lambda}.$$

Therefore, it follows from (2.6) that for any positive x,

$$\mathbb{P}\big(S_n > \|a\|_2\sqrt{2x} + \lambda\|a\|_\infty x\big) \leqslant \exp(-\lambda^2 x).$$

In Section 2.7, we will give a more efficient inequality for sums of random variables with exponential distributions.

2.1.2 Two-sided inequalities

In this subsection, we provide two-sided versions of Bernstein's inequality. The main difference between the two-sided version and the previous one-sided version lies in the assumption on the random variables. More precisely, we shall consider a condition involving the algebraic moments of the random variables.

Theorem 2.9. *Let $X_1,\ldots,X_n$ be a finite sequence of independent random variables with finite algebraic moments at any order. Let S_n, $\mathscr{V}_n$, and v_n be defined as in (2.1). Assume that $\mathbb{E}[S_n] = 0$ and that there exists some positive constant c such that, for any integer $p \geqslant 3$,*

$$\left|\sum_{k=1}^{n}\mathbb{E}[X_k^p]\right| \leqslant \frac{p!\,c^{p-2}}{2}\,\mathscr{V}_n. \tag{2.18}$$

Then, for any positive x,

$$\mathbb{P}(|S_n| \geqslant nx) \leqslant 2\left(1 + \frac{x^2}{2(v_n + cx)}\right)^n \exp\left(-\frac{nx^2}{v_n + cx}\right) \tag{2.19}$$

$$\leqslant 2\exp\left(-\frac{nx^2}{2(v_n + cx)}\right). \tag{2.20}$$

In addition, we also have, for any positive x,

$$\mathbb{P}\big(|S_n| > n(cx + \sqrt{2v_n x})\big) \leqslant 2\exp(-nx). \tag{2.21}$$

Remark 2.10. Obviously, the one-sided bounds for S_n and $-S_n$ hold with the constant one instead of the constant two. One can observe that Condition (2.18) slightly differs from condition (2.2) applied to the two random finite sequences $X_1,\ldots,X_n$ and $-X_1,\ldots,-X_n$. To be more precise, (2.18) is weaker than the two-sided version of (2.2) for odd integers p and stronger for even integers p.

Proof. For any integer $p \geqslant 2$, denote

$$A_{p,n} = \mathbb{E}[X_1^p] + \cdots + \mathbb{E}[X_n^p]. \tag{2.22}$$

It follows from condition (2.18) that for any real t such that $|t|c < 1$,

$$\sum_{k=1}^{n} \mathbb{E}[\cosh(tX_k)] = \sum_{k=1}^{n}\sum_{p=0}^{\infty} \frac{1}{(2p)!}\mathbb{E}[(tX_k)^{2p}] = n + \sum_{p=1}^{\infty} \frac{A_{2p,n}}{(2p)!} t^{2p},$$

$$\leqslant n + \frac{A_{2,n}}{2}\frac{t^2}{1 - c^2 t^2} < \infty.$$

Consequently the random variables $|X_1|, \ldots, |X_n|$ have finite Laplace transforms on the interval $[0, 1/c[$. As $\mathbb{E}[S_n] = 0$, it ensures that for any real t such that $|t|c < 1$,

$$\sum_{k=1}^{n} \mathbb{E}[\exp(tX_k)] = n + \sum_{p=2}^{\infty} \frac{A_{p,n}}{p!} t^p. \tag{2.23}$$

Furthermore, as $A_{2,n} = \mathscr{V}_n$, we can deduce from condition (2.18) that for any real t such that $|t|c < 1$,

$$\left| \sum_{p=2}^{\infty} \frac{A_{p,n}}{p!} t^p \right| \leqslant \mathscr{V}_n \frac{t^2}{1 - |tc|}. \tag{2.24}$$

Therefore, we obtain (2.23) and (2.24) and that for any real t such that $|t|c < 1$,

$$\log\left(\frac{1}{n}\sum_{k=1}^{n} \mathbb{E}[\exp(tX_k)]\right) \leqslant \log\left(1 + \frac{v_n t^2}{2(1 - |tc|)}\right) \leqslant \frac{v_n t^2}{2(1 - |tc|)}.$$

Finally, the rest of the proof is left to the reader inasmuch as it follows the same lines as the proof of Theorem 2.1. $\qquad\square$

2.1.3 About the second term in Bernstein's inequality

In this subsection, we are interested in the second term in Bernstein's type inequalities. Recall that under assumption (2.2), the first improvement of Bennett is given, for any positive x, by

$$\mathbb{P}(S_n \geqslant nx) \leqslant \exp\left(-n g_n(x)\right) \quad \text{where} \quad g_n(x) = \frac{x^2}{v_n + cx + \sqrt{v_n^2 + 2c v_n x}}. \tag{2.25}$$

As x goes to zero, the rate function g_n has the asymptotic expansion

$$g_n(x) = \frac{x^2}{2v_n + 2cx + O(x^2)} = \frac{x^2}{2v_n} - \frac{cx^3}{2v_n^2} + O(x^4).$$

Consequently, the two first terms in the expansion of g_n into powers of x are the same as in the initial inequality of Bernstein. We shall now provide an inequality with a smaller second term.

Theorem 2.11. *Let $X_1, \ldots, X_n$ be a finite sequence of independent random variables with finite variances. Let S_n and v_n be defined as in (2.1). Assume that $\mathbb{E}[S_n] = 0$ and suppose that there exists some positive constant c such that*

$$a_n = \sup_{p \geqslant 3} \left(c^{3-p} \frac{2}{p!} \frac{1}{n} \sum_{k=1}^{n} \mathbb{E}\left[(\max(0, X_k))^p \right] \right) < \infty. \tag{2.26}$$

For any positive constant c satisfying the above condition, denote

$$f_n(x) = \frac{x^2}{v_n + cx + \sqrt{(v_n - cx)^2 + 4a_n x}}. \tag{2.27}$$

Then, for any positive x,

$$\mathbb{P}(S_n \geqslant nx) \leqslant \left(1 + f_n(x) \right)^n \exp\left(-2n f_n(x) \right) \leqslant \exp\left(-n f_n(x) \right). \tag{2.28}$$

Remark 2.12. Under assumption (2.2), we clearly have $a_n \leqslant cv_n$. Then, for any positive x, $\sqrt{(v_n - cx)^2 + 4a_n x} \leqslant v_n + cx$, which immediately leads to

$$f_n(x) \geqslant \frac{x^2}{2(v_n + cx)}.$$

Hence, (2.28) is sharper than the first improvement of Bennett (2.25). Let us now give the asymptotic expansion of f_n. As x goes to zero, it follows from straightforward calculation that

$$f_n(x) = \frac{x^2}{2v_n + 2(a_n/v_n)x + O(x^2)} = \frac{x^2}{2v_n} - \frac{a_n x^3}{2v_n^3} + O(x^4).$$

Consequently, under assumption (2.2), the second term in the expansion of f_n is greater than the second term in the expansion of g_n, which also shows that (2.28) is sharper than (2.25).

Remark 2.13. Denote

$$\alpha_n = \frac{1}{3n} \sum_{k=1}^{n} \mathbb{E}\left[(\max(0, X_k))^3 \right].$$

Obviously $a_n \geqslant \alpha_n$. Now, under the assumptions of Theorem 2.11, $a_n = \alpha_n$ if c is large enough. Hence it seems convenient to choose the smallest c such that $a_n = \alpha_n$ in Theorem 2.11.

Remark 2.14. The function f_n can easily be compared with the rate function g_n given in (2.25). By definition of f_n, $f_n(x) > g_n(x)$ iff $(v_n - cx)^2 + 4a_n x < v_n^2 + 2cv_n x$, which is equivalent to $x < 4(cv_n - a_n)$. Under this condition, the right-hand side term in (2.28) is smaller than the right-hand side term in (2.25).

Proof. We deduce from (2.13) and the definition of a_n that for any positive t such that $0 < tc < 1$,

$$\frac{1}{n}\sum_{k=1}^{n}\mathbb{E}[\exp(tX_k)] \leq 1 + \frac{v_n}{2}t^2 + \frac{a_n}{2(1-tc)}t^3.$$

Therefore, we find from Lemma 2.6 that for any positive t such that $0 < tc < 1$,

$$\frac{1}{n}\log\mathbb{E}[\exp(tS_n)] \leq \log(1 + \psi_n(t)) \quad \text{where} \quad \psi_n(t) = \frac{v_n}{2}t^2 + \frac{a_n}{2(1-tc)}t^3. \quad (2.29)$$

Hence, we immediately obtain from (2.8) and the above upper bound that, for any positive x and for any positive t such that $0 < tc < 1$,

$$\mathbb{P}(S_n \geq nx) \leq (1 + \psi_n(t))^n \exp(-ntx). \quad (2.30)$$

Hereafter, we can choose t in such a way that $\psi_n(t) = tx/2$, which is equivalent to the second order equation

$$(v_nc - a_n)t^2 - (v_n + cx)t + x = 0.$$

If $a_n = v_nc$, then the above equation has a unique solution $t = x/(v_n + cx)$, which clearly belongs to the interval $]0, 1/c[$. In that case, $(v_n - cx)^2 + 4a_nx = (v_n + cx)^2$ and $f_n(x) = tx/2$. Otherwise, the above equation has two real solutions. The solution which lies in the interval $]0, 1/c[$ is

$$t = \frac{v_n + cx - \sqrt{(v_n - cx)^2 + 4a_nx}}{2(v_nc - a_n)} = \frac{2x}{v_n + cx + \sqrt{(v_n - cx)^2 + 4a_nx}}.$$

Hence, once again, we find that $f_n(x) = tx/2$. Finally, we obtain (2.28) by choosing this value of t in (2.30). $\qquad\square$

Example 2.15. Let $\varepsilon_1, \ldots, \varepsilon_n$ be a finite sequence of independent random variables sharing the same Exponential $\mathscr{E}(1)$ distribution. In addition, let $B_1, \ldots, B_n$ be a finite sequence of independent random variables with Bernoulli $\mathscr{B}(1/4)$ distribution. Assume that these two sequences are mutually independent. For all $1 \leq k \leq n$, denote

$$X_k = B_k\varepsilon_k - \frac{1}{6}\left(1 - B_k\right)\varepsilon_k^2.$$

One can easily check that, for all $1 \leq k \leq n$, $\mathbb{E}[X_k] = 0$, $\mathbb{E}[X_k^2] = 1 = v_n$, and $\mathbb{E}[(\max(0, X_k))^p] = p!/4$. Taking $c = 1$, we obtain from the definition (2.26) of a_n that $a_n = 1/2$. Then, it follows from Theorem 2.11 that, for any positive x,

$$\mathbb{P}(S_n \geq nx) \leq (1 + f_n(x))^n \exp(-2nf_n(x))$$

where

$$f_n(x) = \frac{x^2}{1 + x + \sqrt{1 + x^2}}.$$

Figure 2.3 below compares the rate function

$$\Psi(x) = \frac{2x^2}{1 + x + \sqrt{1 + x^2}} - \log\left(1 + \frac{x^2}{1 + x + \sqrt{1 + x^2}}\right)$$

appearing here with the two rate functions of Figure 2.1. One can realize in this example that inequality (2.28) is more efficient than (2.5) and (2.3).

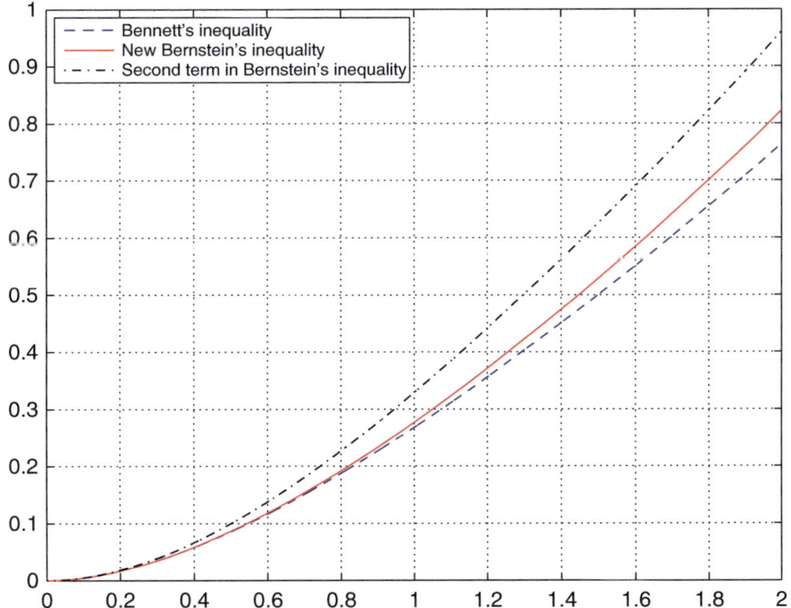

Fig. 2.3 Comparisons in Bernstein's inequalities with the new second term

2.2 Hoeffding's inequality

In this section, we focus our attention on the classical Hoeffding's inequality [11], which requires that the random variables are bounded from above and from below. We shall also establish Antonov's type extensions of this inequality. Let us start with the classical inequality of Hoeffding.

Theorem 2.16 (Hoeffding's inequality). *Let* $X_1, \ldots, X_n$ *be a finite sequence of independent random variables. Assume that for all* $1 \leqslant k \leqslant n$, *one can find*

two constants $a_k < b_k$ such that $a_k \leqslant X_k \leqslant b_k$ almost surely. Denote $S_n = X_1 + \ldots + X_n$. Then, for any positive x,

$$\mathbb{P}(|S_n - \mathbb{E}[S_n]| \geqslant x) \leqslant 2\exp\left(-\frac{2x^2}{D_n}\right) \quad \text{where} \quad D_n = \sum_{k=1}^{n}(b_k - a_k)^2. \tag{2.31}$$

Remark 2.17. We shall see in the proof of Theorem 2.16 below that $D_n \geqslant 4V_n$ where $V_n = \mathrm{Var}(S_n)$. Clearly, the equality $D_n = 4V_n$ does not hold true, unless the random variables $X_1, \ldots, X_n$ have the Bernoulli distribution given, for all $1 \leqslant k \leqslant n$, by $\mathbb{P}(X_k = a_k) = \mathbb{P}(X_k = b_k) = 1/2$. In Section 2.5, we will establish a more efficient inequality. To be more precise, we will improve Hoeffding's inequality (2.31) replacing D_n by the lower bound given by the inequality

$$D_n \geqslant \frac{2}{3}D_n + \frac{4}{3}V_n.$$

The proof of Hoeffding's inequality relies on the following lemmas which give upper bounds for the variances and the Laplace transforms of the random variables $X_1, \ldots, X_n$.

Lemma 2.18. *Let X be a random variable with finite variance σ^2. Assume that $a \leqslant X \leqslant b$ almost surely for some real constants a and b. Denote $m = \mathbb{E}[X]$. Then,*

$$\sigma^2 \leqslant (b-m)(m-a) \leqslant \frac{(b-a)^2}{4}. \tag{2.32}$$

Proof. The convexity of the square function implies that $X^2 \leqslant (a+b)X - ab$ almost surely. Hence $\sigma^2 = \mathbb{E}[X^2] - m^2 \leqslant (a+b)m - ab - m^2 \leqslant -ab + (a+b)^2/4$, which implies the lemma. $\square$

Lemma 2.19. *Let X be a random variable with finite variance σ^2. Assume that $a \leqslant X \leqslant b$ almost surely for some real constants a and b. Then, for any real t,*

$$\log\left(\mathbb{E}[\exp(tX)]\right) \leqslant t\mathbb{E}[X] + \frac{t^2}{8}(b-a)^2.$$

Proof. Let L and ℓ be the Laplace and log-Laplace transforms of X. As the random variable X is bounded from above and from below, L and ℓ are real analytic functions. Moreover, for any real t, $\ell(t) = \log L(t)$,

$$\ell'(t) = \frac{L'(t)}{L(t)} \quad \text{and} \quad \ell''(t) = \frac{L''(t)}{L(t)} - \left(\frac{L'(t)}{L(t)}\right)^2.$$

Consider the classical change of probability

$$\frac{d\mathbb{P}_t}{d\mathbb{P}} = \exp\left(tX - \ell(t)\right) = \frac{\exp(tX)}{L(t)}$$

and denote by $\mathbb{E}_t$ the expectation associated with $\mathbb{P}_t$. One can observe that for any integrable random variable Y,

$$\mathbb{E}_t[Y] = \frac{\mathbb{E}[Y\exp(tX)]}{L(t)}.$$

In particular,

$$E_t[X] = \frac{\mathbb{E}[X\exp(tX)]}{L(t)} = \frac{L'(t)}{L(t)}, \qquad E_t[X^2] = \frac{\mathbb{E}[X^2\exp(tX)]}{L(t)} = \frac{L''(t)}{L(t)}.$$

Consequently, $\ell''(t) = \mathbb{E}_t[X^2] - \mathbb{E}_t^2[X]$, which means that $\ell''(t)$ is equal to the variance of the random variable X under the new probability $\mathbb{P}_t$. As the random variable X takes its values in $[a,b]$ almost surely, we may apply Lemma 2.18 under the new probability $\mathbb{P}_t$, which gives $\ell''(t) \leqslant (b-a)^2/4$. Since $\ell(0) = 0$ and $\ell'(0) = \mathbb{E}[X]$, it completes the proof of Lemma 2.19. $\qquad\square$

Proof of Theorem 2.16. We shall now proceed to the proof of Hoeffding's inequality. We deduce from Lemma 2.19 together with the independence of the random variables $X_1,\ldots,X_n$ that, for any real t,

$$\log\mathbb{E}\left[\exp(tS_n)\right] = \sum_{k=1}^{n}\log\mathbb{E}\left[\exp(tX_k)\right] \leqslant t\mathbb{E}[S_n] + \frac{t^2}{8}D_n \qquad (2.33)$$

where D_n is given by (2.31). For any positive t, it follows from Markov's inequality applied to $\exp(tS_n)$ that

$$\log\mathbb{P}(S_n \geqslant \mathbb{E}[S_n] + x) \leqslant -tx - t\mathbb{E}[S_n] + \log\mathbb{E}[\exp(tS_n)]. \qquad (2.34)$$

Consequently, inequalities (2.34) and (2.33) imply that for all $x \geqslant 0$ and $t > 0$,

$$\mathbb{P}(S_n - \mathbb{E}[S_n] \geqslant x) \leqslant \exp\left(-tx + \frac{t^2}{8}D_n\right).$$

By taking the optimal value $t = 4x/D_n$, we find that

$$\mathbb{P}(S_n - \mathbb{E}[S_n] \geqslant x) \leqslant \exp\left(-\frac{2x^2}{D_n}\right). \qquad (2.35)$$

Replacing X_k by $-X_k$, we obtain by the same token that for all $x \geqslant 0$,

$$\mathbb{P}(S_n - \mathbb{E}[S_n] \leqslant -x) \leqslant \exp\left(-\frac{2x^2}{D_n}\right). \qquad (2.36)$$

Therefore, (2.31) follows from (2.35) and (2.36), which completes the proof of Theorem 2.16. $\qquad\square$

It is necessary to add some comments on Hoeffding's inequality. Recall that for all $1 \leqslant k \leqslant n$, $a_k \leqslant X_k \leqslant b_k$ almost surely, where $a_k < b_k$. For all $1 \leqslant k \leqslant n$, let Z_k be the random variable defined by $Z_k = (X_k - a_k)/c_k$ with $c_k = b_k - a_k$, which is equivalent to $X_k = a_k + c_k Z_k$. For all $1 \leqslant k \leqslant n$, the random variable Z_k takes its values in the interval $[0,1]$ and

$$S_n - \mathbb{E}[S_n] = \sum_{k=1}^{n} c_k(Z_k - \mathbb{E}[Z_k]). \tag{2.37}$$

Consequently $S_n - \mathbb{E}[S_n]$ is a weighted sum of independent random variables whose laws have a support included in an interval of length 1. Hence, the support of the law of S_n is included in an interval of length $\|c\|_1 = c_1 + \cdots + c_n$ and

$$\mathbb{P}(|S_n - \mathbb{E}[S_n]| > \|c\|_1) = 0, \tag{2.38}$$

which cannot be deduced from Hoeffding's inequality. Furthermore, for any $p \geqslant 1$, denote

$$\|c\|_p = \left(\sum_{k=1}^{n} c_k^p \right)^{1/p}.$$

Hoeffding's inequality (2.31) is clearly equivalent, for any positive x, to

$$\mathbb{P}(|S_n - \mathbb{E}[S_n]| \geqslant \|c\|_2 x) \leqslant 2\exp(-2x^2). \tag{2.39}$$

Antonov [1] extended inequality (2.39) by proving that for any p in $]1,2]$ and for any positive x,

$$\mathbb{P}(|S_n - \mathbb{E}[S_n]| \geqslant \|c\|_p x) \leqslant 2\exp(-C_q x^q) \tag{2.40}$$

where $q = p/(p-1)$ and the constant C_q obtained by Antonov converges to 0 as q tends to ∞. Observe, however, that Antonov [1] also proved inequality (2.40) under the more general tail assumption that, for all $1 \leqslant k \leqslant n$, $\mathbb{P}(|Z_k| \geqslant x) \leqslant \alpha\exp(-\beta x^q)$ for some positive constants α and β. Here, we will prove that the constant C_q appearing in (2.40) is larger than 2. We refer the reader to Rio [21] for more details about the constants in (2.40).

Theorem 2.20. *Let $X_1, \ldots, X_n$ be a finite sequence of independent random variables. Assume that for all $1 \leqslant k \leqslant n$, one can find two constants $a_k < b_k$ such that $a_k \leqslant X_k \leqslant b_k$ almost surely and let $c_k = b_k - a_k$. Denote $S_n = X_1 + \cdots + X_n$. Then, for any p in $]1,2]$ and for any positive x,*

$$\mathbb{P}(|S_n - \mathbb{E}[S_n]| \geqslant \|c\|_p x) \leqslant 2\exp(-2x^q) \tag{2.41}$$

where $q = p/(p-1)$.

Remark 2.21. Let x be any real in $]1,\infty[$. Then, by taking the limit as q tends to infinity in (2.41), we obtain (2.38).

The proof of Theorem 2.20 relies on the following lemma on convex functions.

Lemma 2.22. *Let ℓ be a convex and increasing function from $[0,\infty[$ to $[0,\infty]$ such that $\ell(0)=0$ and $\ell'(0)=0$. Denote by ℓ^* the Fenchel-Legendre transform of ℓ. Then, for any real p in $]1,2]$, we have*

$$a_\ell(p) = \sup_{t>0}\big(t^{-p}\ell(t)\big) = p^{-1}\big(q\inf_{x>0}\big(x^{-q}\ell^*(x)\big)\big)^{1-p} \text{ where } q = p/(p-1). \quad (2.42)$$

Proof. Since ℓ is a convex function $\ell = (\ell^*)^*$. Hence, for any positive t,

$$\ell(t) = \sup_{x>0}\big(xt - \ell^*(x)\big),$$

which implies that

$$a_\ell(p) = \sup_{x>0}\big(\sup_{t>0} t^{-p}\big(xt - \ell^*(x)\big)\big). \quad (2.43)$$

For any positive t, denote $f(t) = t^{-p}(xt - \ell^*(x))$. In order to prove (2.42), it is necessary to compute the maximum of the function f. From the assumption $\ell(0) = \ell'(0) = 0$, we know that $\ell^*(x) > 0$ for any positive x. Moreover, we also have

$$f'(t) = t^{-1-p}\big(p\ell^*(x) - (p-1)xt\big).$$

Hence, f has a unique maximum at the point $t_x = q\ell^*(x)/x$ where $q = p/(p-1)$. Therefore,

$$\sup_{t>0} t^{-p}\big(xt - \ell^*(x)\big) = (q-1)\ell^*(x)t_x^{-p} = (q-1)q^{-p}\big(x^q/\ell^*(x)\big)^{p-1}. \quad (2.44)$$

Consequently, we deduce from (2.43) and (2.44) that

$$a_\ell(p) = \frac{q-1}{q}\sup_{x>0}\Big(\frac{x^q}{q\ell^*(x)}\Big)^{p-1},$$

which implies Lemma 2.22. $\qquad\square$

Proof of Theorem 2.20. Let Z be a centered random variable with values in the interval $[a, a+1]$ with $a < 0$. It follows from Lemma 2.19 that for any real t,

$$\ell_Z(t) = \log\mathbb{E}[\exp(tZ)] \leqslant \frac{t^2}{8}.$$

Moreover, as $a < 0$, $Z \leqslant 1$ almost surely, which implies that

$$\ell_Z'(t) = \frac{\mathbb{E}[Z\exp(tZ)]}{\mathbb{E}[\exp(tZ)]} \leqslant 1.$$

From the two above inequalities and the convexity of the log-Laplace transform ℓ_Z, we obtain that, for any real t,

$$\ell_Z(t) \leqslant \ell(t), \quad (2.45)$$

where $\ell(t) = t^2/8$ if $t \leqslant 4$ and $\ell(t) = t - 2$ if $t \geqslant 4$. Hereafter, starting from (2.37) and applying (2.45) to the random variables $Z_k - \mathbb{E}[Z_k]$, we find that for any real t,

$$\log \mathbb{E}[\exp(tS_n)] - t\mathbb{E}[S_n] \leqslant \sum_{k=1}^{n} \ell(c_k t) \leqslant \sup_{x>0}\left(x^{-p}\ell(x)\right) \|c\|_p^p t^p. \tag{2.46}$$

Denote by ℓ^* the Legendre-Fenchel transform of the convex function ℓ. It follows from straightforward calculation that $\ell^*(x) = 2x^2$ for x in $[0,1]$ and $\ell_0^*(x) = +\infty$ for $x > 1$. Hence, for any $q \geqslant 2$,

$$\inf_{x>0} x^{-q}\ell_0^*(x) = 2.$$

Consequently, we obtain from (2.46) and Lemma 2.22 that, for any real t,

$$\log \mathbb{E}[\exp(tS_n)] - t\mathbb{E}[S_n] \leqslant p^{-1}(2q)^{1-p}\|c\|_p^p t^p. \tag{2.47}$$

Hence, we deduce from (2.47) together with Markov's inequality that for all $x \geqslant 0$ and $t > 0$,

$$\mathbb{P}(S_n - \mathbb{E}[S_n] \geqslant \|c\|_p x) \leqslant \exp\left(tx - (2q)^{1-p}(t^p/p)\right).$$

By taking the optimal value $t = 2qx^{q-1}$ in this inequality, we find that

$$\mathbb{P}(S_n - \mathbb{E}[S_n] \geqslant \|c\|_p x) \leqslant \exp\left(-2x^q\right). \tag{2.48}$$

Replacing X_k by $-X_k$, we obviously obtain the same inequality, which completes the proof of Theorem 2.20. $\qquad\qquad\qquad\qquad\qquad\qquad\qquad\qquad\qquad\qquad\quad\square$

2.3 Binomial rate functions

Throughout this section, we assume that $X_1, \ldots, X_n$ is a finite sequence of independent random variables bounded from above. More precisely, we assume that there exists a positive constant b such that, for all $1 \leqslant k \leqslant n$,

$$X_k \leqslant b \qquad \text{a.s.} \tag{2.49}$$

The purpose of this section is to compare the tails of their sum S_n on the right with the tails of binomial random variables. The fundamental tool of the proofs is the lemma below, whose second part is due to Bennett [2]. We refer to Bentkus [4] for the first part of the lemma and for other results in this direction under condition (2.49), and to Pinelis [19] for a refinement of (2.49) under some additional condition on the random variable X appearing in the lemma below. Next we will deduce Theorems 1 and 3 of Hoeffding [11] from this key lemma.

Lemma 2.23. *Let X be a centered random variable with finite variance σ^2. Assume that $X \leqslant b$ almost surely for some positive real b. Let v be any positive real such that $\sigma^2 \leqslant v$. Denote by ξ the Bernoulli type random variable with mean 0 and variance v defined by*

$$\mathbb{P}(\xi = b) = \frac{v}{b^2 + v} \qquad and \qquad \mathbb{P}(\xi = -v/b) = \frac{b^2}{b^2 + v}.$$

Then, for any real t,

$$\mathbb{E}[(X - t)_+^2] \leqslant \mathbb{E}[(\xi - t)_+^2]. \tag{2.50}$$

Consequently, for any positive real t,

$$\mathbb{E}[\exp(tX)] \leqslant \mathbb{E}[\exp(t\xi)] = \frac{v}{b^2 + v} \exp(tb) + \frac{b^2}{b^2 + v} \exp(-tv/b). \tag{2.51}$$

Proof. If $t \geqslant b$, then the two expectations in (2.50) vanish and there is nothing to prove. If $t \leqslant -v/b$, then

$$\mathbb{E}[(X - t)_+^2] \leqslant \mathbb{E}[(X - t)^2] = \sigma^2 + t^2 \leqslant v + t^2 = \mathbb{E}[(\xi - t)^2] = \mathbb{E}[(\xi - t)_+^2].$$

If t belongs to $[-v/b, b]$, then for any $x \in [-v/b, b]$,

$$(x - t)_+ \leqslant \frac{b - t}{b^2 + v}(bx + v)_+$$

leading to

$$\mathbb{E}[(X - t)_+^2] \leqslant \frac{(b - t)^2}{(b^2 + v)^2} \mathbb{E}[(bX + v)_+^2] \leqslant \frac{(b - t)^2}{(b^2 + v)^2} \mathbb{E}[(bX + v)^2].$$

However,

$$\frac{(b - t)^2}{(b^2 + v)^2} \mathbb{E}[(bX + v)^2] = \frac{(b - t)^2 v}{b^2 + v} = \mathbb{E}[(\xi - t)_+^2],$$

which shows that (2.50) still holds. We are now in position to prove (2.51). For any $x \in \mathbb{R}$ and any positive t, we have the integral representation

$$\mathbb{E}[\exp(tX)] = \frac{t^2}{2} \int_{\mathbb{R}} \mathbb{E}\left[\left(X - \frac{s}{t}\right)_+^2\right] \exp(s) ds.$$

Consequently, (2.51) immediately follows from (2.50), which completes the proof of Lemma 2.23. $\qquad\square$

We now apply (2.51) to sums of independent random variables $X_1, \ldots, X_n$ bounded from above. In the case of centered random variables, the result below coincides with Theorem 3 in Hoeffding [11]. We also refer to Bennett [3] for an analogous bound under the additional assumption that the random variables $X_1, \ldots, X_n$ are symmetrically distributed about their mean.

Theorem 2.24. *Let $X_1, \ldots, X_n$ be a finite sequence of independent random variables with finite variances satisfying (2.49) for some positive real b. Let S_n and v_n be defined as in (2.1) and assume that $\mathbb{E}[S_n] = 0$. Then, for any $v \geqslant v_n$ and for any x in $[0, b]$,*

$$\mathbb{P}(S_n \geqslant nx) \leqslant \exp\left(-n\left(\left(\frac{v+bx}{v+b^2}\right)\log\left(1+\frac{bx}{v}\right) + \left(\frac{b^2-bx}{b^2+v}\right)\log\left(1-\frac{x}{b}\right)\right)\right),$$

$$\leqslant \exp\left(-ng(b,v)x^2\right), \tag{2.52}$$

where

$$g(b,v) = \begin{cases} \dfrac{b^2}{(b^4-v^2)}\log\left(\dfrac{b^2}{v}\right) & \text{if } v < b^2, \\[4mm] \dfrac{1}{2v} & \text{if } v \geqslant b^2. \end{cases} \tag{2.53}$$

Remark 2.25. Note that $S_n \leqslant nb$ almost surely, which implies that $\mathbb{P}(S_n > nb) = 0$.

Proof. We shall only prove Theorem 2.24 in the particular case $b = 1$, inasmuch as the general case follows by dividing the initial random variables by b. According to Lemma 2.6, we have for any real t

$$\frac{1}{n}\log\mathbb{E}[\exp(tS_n)] \leqslant \ell(t) \quad \text{where} \quad \ell(t) = \log\left(\frac{1}{n}\sum_{k=1}^{n}\mathbb{E}[\exp(tX_k)]\right).$$

Denote by X a random variable with distribution

$$\mu = \frac{1}{n}(\mu_1 + \cdots + \mu_n)$$

where, for all $1 \leqslant k \leqslant n$, μ_k stands for the distribution of X_k. We clearly have $X \leqslant 1$ almost surely. In addition, $\mathbb{E}[X] = 0$ and

$$\mathbb{E}[X^2] = \frac{1}{n}(\mathbb{E}[X_1^2] + \cdots + \mathbb{E}[X_n^2]) \leqslant v.$$

Hence, according to (2.51), we have for any positive t,

$$\frac{1}{n}\sum_{k=1}^{n}\mathbb{E}[\exp(tX_k)] = \mathbb{E}[\exp(tX)] \leqslant \frac{v}{1+v}\exp(t) + \frac{1}{1+v}\exp(-vt).$$

It implies that, for any positive t,

$$\log\mathbb{E}[\exp(tS_n)] \leqslant nL_v(t) \tag{2.54}$$

where

$$L_v(t) = \log(ve^t + e^{-vt}) - \log(1+v).$$

In order to prove the first part of Theorem 2.24, it remains to prove that the Legendre-Fenchel transform L_v^* of L_v is given by

$$L_v^*(x) = \begin{cases} \left(\dfrac{v+x}{v+1}\right)\log\left(1+\dfrac{x}{v}\right) + \left(\dfrac{1-x}{1+v}\right)\log(1-x) & \text{if } x \in [0,1], \\ +\infty & \text{if } x > 1. \end{cases} \tag{2.55}$$

In order to compute $L_v^*(x)$, we have to solve the equation $L_v'(t) = x$. On the one hand, for $x < 1$,

$$L_v'(t) = \frac{v(1 - e^{-t(1+v)})}{v + e^{-t(1+v)}} = x \iff (1+v)t = \log\left(1 + \frac{x}{v}\right) - \log(1-x).$$

For this value of t,

$$tx - L_v(x) = \left(\frac{v+x}{v+1}\right)\log\left(1 + \frac{x}{v}\right) + \left(\frac{1-x}{1+v}\right)\log(1-x) = L_v^*(x),$$

which gives (2.55) in the case $x < 1$. On the other hand, for $x \geqslant 1$, the function $t \to tx - L_v(t)$ is increasing. Hence, if $x = 1$,

$$L_v^*(x) = \lim_{t \to \infty}(t - L_v(t)) = \log\left(1 + \frac{1}{v}\right)$$

while, if $x > 1$, $L_v^*(x) = +\infty$, which completes the proof of (2.55). We deduce the first part of Theorem 2.24 from (2.54) and (2.55). It now remains to prove the second part of Theorem 2.24. This second part immediately follows from the lower bound below on L_v^*, due to Hoeffding [11]. $\qquad\square$

Lemma 2.26. *Let L_v^* be defined as in* (2.55). *Then, we have*

$$\inf_{x>0}\left(\frac{L_v^*(x)}{x^2}\right) = \begin{cases} \dfrac{|\log(v)|}{1 - v^2} & \text{if } v < 1, \\ \dfrac{1}{2v} & \text{if } v \geqslant 1. \end{cases} \tag{2.56}$$

Proof. Since $L_v^*(x) = +\infty$ for $x > 1$,

$$\inf_{x>0}\left(\frac{L_v^*(x)}{x^2}\right) = \inf_{0<x\leqslant 1}\left(\frac{L_v^*(x)}{x^2}\right).$$

Let ψ_v be the function defined by $\psi_v(x) = x^{-2}L_v^*(x)$ if x in $]0,1]$ and $\psi_v(0) = 1/(2v)$. Then, ψ_v is differentiable on $[0,1[$, and for any x in $[0,1[$,

$$x^2\psi_v'(x) = (L_v^*)'(x) - 2x^{-1}L_v^*(x). \tag{2.57}$$

We already saw that $(1+v)(L_v^*)'(x) = \log(1+x/v) - \log(1-x)$. Hence, if φ_v is the function defined, for all x in $]0,1[$, by $\varphi_v(x) = (1+v)x^2\psi_v'(x)$, it follows that

$$\varphi_v(x) = \left(1 - \frac{2}{x}\right)\log(1-x) - \left(1 + \frac{2v}{x}\right)\log\left(1 + \frac{x}{v}\right).$$

Apparently it seems difficult to find the roots of the above function. However,

$$\left(1 + \frac{2v}{x}\right)\log\left(1 + \frac{x}{v}\right) = -\left(1 + \frac{2v}{x}\right)\log\left(\frac{v}{v+x}\right),$$

$$= \left(1 - \frac{2(v+x)}{x}\right)\log\left(1 - \frac{x}{v+x}\right).$$

Therefore, one can realize that, for all x in $]0,1[$,

$$\varphi_v(x) = H(x) - H\left(\frac{x}{v+x}\right) \quad \text{where} \quad H(x) = \left(1 - \frac{2}{x}\right)\log(1-x). \qquad (2.58)$$

Furthermore, for all x in $]0,1[$, we have the expansion

$$H(x) = 2 + \sum_{k=2}^{\infty} \frac{(k-1)}{k(k+1)}x^k.$$

In this expansion, the coefficients are positive. Thus, H is increasing on $]0,1[$. Hence, it follows from (2.57) and (2.58) that $\psi_v'(x) > 0$ if and only if $x > x/(v+x)$, that is $x > 1 - v$. Therefrom, if $v \geqslant 1$, then ψ_v is increasing on $]0,1]$. Consequently, the value of the minimum is

$$C_v = \lim_{x \searrow 0} \frac{L_v^*(x)}{x^2} = \frac{1}{2v},$$

which gives the second part of Lemma 2.26. If $v < 1$, then ψ_v has its minimum at $x = 1 - v$ and the value of this minimum is

$$C_v = \psi_v(1-v) = \frac{|\log(v)|}{1 - v^2},$$

which completes the proof of Lemma 2.26. $\qquad\qquad\qquad\qquad\qquad\qquad\qquad\qquad \square$

To conclude this section, we apply Theorem 2.24 to independent random variables with values in $[0,1]$. The result below is exactly Theorem 1 in Hoeffding [11].

Theorem 2.27. *Let $X_1,\ldots,X_n$ be a finite sequence of independent random variables with values in $[0,1]$ and denote $\mu = \mathbb{E}[S_n]/n$. Then, for any x in $]\mu,1[$,*

$$\mathbb{P}(S_n \geqslant nx) \leqslant \exp\left(-n\left(x\log\left(\frac{x}{\mu}\right) + (1-x)\log\left(\frac{1-x}{1-\mu}\right)\right)\right),$$

$$\leqslant \exp\left(-ng(\mu)(x-\mu)^2\right), \qquad (2.59)$$

$$\leqslant \exp\left(-2n(x-\mu)^2\right),$$

where

$$g(\mu) = \begin{cases} \dfrac{1}{1-2\mu}\log\!\left(\dfrac{1-\mu}{\mu}\right) & \text{if } 0 < \mu < \dfrac{1}{2}, \\[2ex] \dfrac{1}{2\mu(1-\mu)} & \text{if } \dfrac{1}{2} \leqslant \mu < 1. \end{cases}$$

Proof. If $\mu = 0$ or $\mu = 1$, then $S_n = n\mu$ almost surely, and there is nothing to prove. Hereafter, assume that $0 < \mu < 1$ and denote, for all $1 \leqslant k \leqslant n$, $Y_k = X_k - \mu$. The random variables $Y_1,\ldots,Y_n$ satisfy the assumptions of Theorem 2.24 with $b = 1-\mu$. Moreover, if $\Sigma_n = Y_1 + \cdots + Y_n$, we clearly have $\mathbb{E}[\Sigma_n] = \mathbb{E}[S_n] - n\mu = 0$. In addition, for all $1 \leqslant k \leqslant n$,

$$\mathbb{E}[Y_k^2] = \mathbb{E}[X_k^2] - 2\mu\mathbb{E}[X_k] + \mu^2 \leqslant (1-2\mu)\mathbb{E}[X_k] + \mu^2,$$

since $\mathbb{E}[X_k^2] \leqslant \mathbb{E}[X_k]$. It leads to

$$\mathrm{Var}(\Sigma_n) = \mathbb{E}[Y_1^2] + \cdots + \mathbb{E}[Y_n^2] \leqslant n\mu(1-2\mu) + n\mu^2 = n\mu(1-\mu).$$

Finally, Theorem 2.27 immediately follows from Theorem 2.24 with $b = 1-\mu$ and $v = \mu(1-\mu)$. $\qquad\square$

2.4 Bennett's inequality

In this section, we deduce Bennett's type inequalities from the results of Section 2.3. First of all, let h and h_w be the functions defined by

$$h(x) = \begin{cases} (1+x)\log(1+x) - x & \text{if } x > -1, \\ 1 & \text{if } x = -1, \\ +\infty & \text{if } x < -1, \end{cases} \tag{2.60}$$

and

$$h_w(x) = \begin{cases} \dfrac{h(wx)}{w^2} & \text{if } w \neq 0, \\[2ex] \dfrac{x^2}{2} & \text{if } w = 0. \end{cases} \tag{2.61}$$

Theorem 2.28. *Let $X_1,\ldots,X_n$ be a finite sequence of independent random variables satisfying (2.49) for some positive constant b. Let S_n and v_n be defined as in (2.1) and assume that $\mathbb{E}[S_n] = 0$. Let $w_n = (b/v_n) - (1/b)$. Then, for any x in $[0,b]$,*

$$\mathbb{P}(S_n \geqslant nx) \leqslant \exp\!\left(-\frac{n}{v_n} h_{w_n}(x)\right), \tag{2.62}$$

$$\leqslant \exp\!\left(-\frac{nv_n}{b^2} h\!\left(\frac{bx}{v_n}\right)\right),$$

where the above functions are given by (2.60) and (2.61). Hence, if $v_n \geqslant b^2$, then, for any positive x,

$$\mathbb{P}(S_n \geqslant nx) \leqslant \exp\left(-\frac{nx^2}{2v_n}\right). \tag{2.63}$$

Furthermore, for any x in $[0,b]$,

$$\mathbb{P}(S_n \geqslant nx) \leqslant \exp\left(-\frac{nx^2}{2(v_n+bx/3)(1-x/(3b))}\right) \tag{2.64}$$

$$\leqslant \exp\left(-\frac{nx^2}{2(v_n+bx/3)}\right).$$

Remark 2.29. The second upper bound in (2.62) is known as Bennett's inequality. This inequality was called second improvement of Bernstein's inequality in Bennett [2]. This second improvement is more efficient than the first improvement of Bennett, which corresponds to inequality (2.4) with $c = b/3$. We now discuss the first upper bound. Elementary computations show that h_w are decreasing with respect to w. Hence, the first upper bound is more efficient than Bennett's inequality. For example, if $v_n \geqslant b^2$, then $w_n \leqslant 0$. In that case, $h_{w_n}(x) \geqslant x^2/2$, which implies (2.63). On the other side, (2.63) cannot be deduced from Bennett's inequality. Note, however, that (2.63) has already been established in Theorem 2.24 of Section 2.3. It comes from the proof of Theorem 2.28 that the above upper bounds are increasing with respect to v_n. Consequently, one can replace v_n by any real $v \geqslant v_n$ in Theorem 2.28.

Remark 2.30. Inequality (2.64) is an improved version of Bernstein's inequality for bounded random variables, which implies (2.63). Figure 2.4 below compares the rate functions given in Theorem 2.28 in the particular case $b = 1$ and $v_n = 1$. For any x in $[0,1]$,

$$\Phi(x) = \frac{x^2}{2(1+x/3)} \quad \text{and} \quad \Psi(x) = \frac{x^2}{2(1+x/3)(1-x/3)}$$

are the rate functions in Bernstein's inequality and its improvement for bounded random variables, respectively. In addition, for any x in $[0,1]$,

$$h(x) = (1+x)\log(1+x) - x, \qquad \ell(x) = \frac{x^2}{2},$$

$$\varphi(x) = \frac{1}{2}(1+x)\log(1+x) + \frac{1}{2}(1-x)\log(1-x)$$

are the rate functions in Bennett's inequality, the improvement of Bennett's inequality, and the Binomial rate function. One can realize that inequality (2.52) with the Binomial rate function outperforms Bernstein and Bennett inequalities.

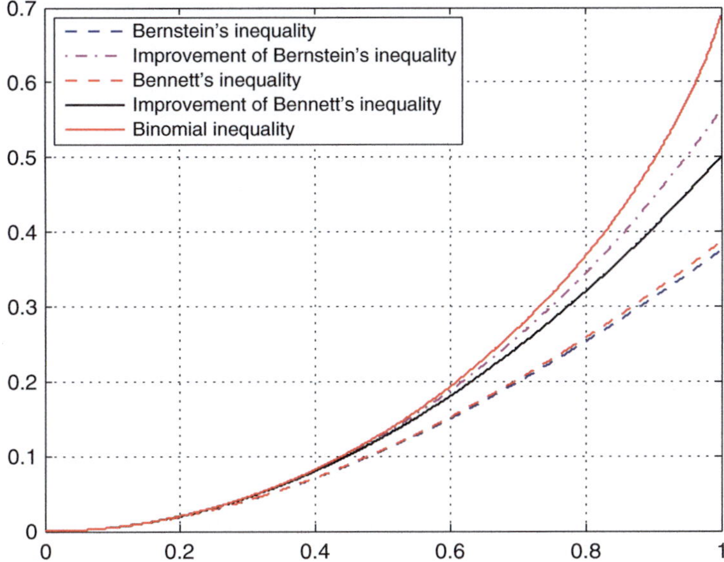

Fig. 2.4 Comparisons in Bennett's inequalities

Remark 2.31. An interesting situation occurs when w_n is different from zero and v_n is small. Figure 2.5 below compares the more efficient rate functions of Theorem 2.28 given, for any x in $[0,1]$, by

$$\Psi_n(x) = \frac{x^2}{2(v_n + bx/3)(1 - x/(3b_n))},$$

$$h_n(x) = \frac{1}{v_n w_n^2}\Big((1 + w_n x)\log(1 + w_n x) - w_n x\Big),$$

and

$$\varphi_n(x) = \left(\frac{v_n + x}{v_n + 1}\right)\log\left(1 + \frac{x}{v_n}\right) + \left(\frac{1-x}{1+v_n}\right)\log(1 - x)$$

in the special case $b = 1$ and $v_n = 1/20$. One can observe that inequality (2.52) with the Binomial rate function still outperforms the improvements of Bernstein and Bennett inequalities. However, the improvement of Bennett's inequality behaves better than the improvement of Bernstein's inequality.

Proof. The fact that (2.62) implies (2.63) is already proven in Remark 2.29. We shall only prove Theorem 2.28 in the special case $b = 1$, inasmuch as the general case follows by dividing the initial random variables by b. Throughout the proof, $v = v_n$ and $w = w_n = (1 - v)/v$. According to (2.52), we have for any x in $[0,1]$,

$$\mathbb{P}(S_n \geqslant nx) \leqslant \exp(-nL_v^*(x))$$

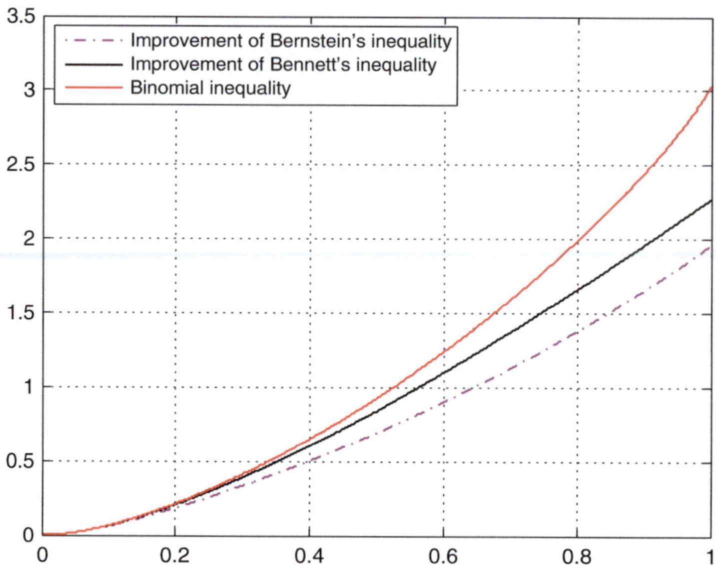

Fig. 2.5 Comparisons in the situation where w_n is different from zero

where

$$L_v^*(x) = \left(\frac{v+x}{v+1}\right) \log\left(1 + \frac{x}{v}\right) + \left(\frac{1-x}{1+v}\right) \log(1-x)$$

with the convention $0\log 0 = 0$. Furthermore, $\mathbb{P}(S_n > n) = 0$. Consequently (2.62) and (2.64) follow from the lower bounds below on L_v^*, which achieves the proof of Theorem 2.28. $\qquad\qquad\square$

Lemma 2.32. *For any x in $[0,1]$, we have*

$$L_v^*(x) \geqslant \frac{h_w(x)}{v} \geqslant vh\left(\frac{x}{v}\right)$$

where $w = (1-v)/v$. Moreover, for any x in $[0,1]$,

$$L_v^*(x) \geqslant \frac{x^2}{2(v+x/3)(1-x/3)}.$$

Proof. In order to prove the first part of Lemma 2.32, one can observe that $L_v^*(0) = (L_v^*)'(0) = 0$ and

$$(L_v^*)''(x) = \frac{1}{(v+x)(1-x)}.$$

Consequently, for any x in $[0,1]$,

$$L_v^*(x) = \int_0^x (x-t)(L_v^*)''(t)dt = \int_0^x \frac{(x-t)}{(v+t)(1-t)}dt.$$

For any t in $[0,1[$, $0 < (v+t)(1-t) \leqslant v+(1-v)t \leqslant v+t$. Hence,

$$L_v^*(x) \geqslant \int_0^x \frac{x-t}{v+(1-v)t}\,dt \geqslant \int_0^x \frac{x-t}{v+t}\,dt. \tag{2.65}$$

Now,

$$\int_0^x \frac{x-t}{v+(1-v)t}\,dt = \frac{h_w(x)}{v} \qquad \text{and} \qquad \int_0^x \frac{x-t}{v+t}\,dt = vh\left(\frac{x}{v}\right),$$

which completes the proof of the first part. The main tool of the proof of the second part is an expansion of L_v^* into power series. We start by noting that

$$(1+v)L_v^*(x) = vh\left(\frac{x}{v}\right) + h(-x). \tag{2.66}$$

Next, for any x in $[0,1]$,

$$h(-x) = \frac{x^2}{2} \sum_{k=0}^{\infty} \frac{2x^k}{(k+2)(k+1)}. \tag{2.67}$$

In order to give an expansion of $vh(x/v)$, we note that, for any positive x,

$$h(x) = -(1+x)\log\left(1 - \frac{x}{1+x}\right) - x - \sum_{k=2}^{\infty} \frac{1}{k}\frac{x^k}{(1+x)^{k-1}},$$

which leads to the expansion

$$vh\left(\frac{x}{v}\right) = \frac{x^2}{2(v+x)} \sum_{k=0}^{\infty} \frac{2}{k+2}\left(\frac{x}{v+x}\right)^k. \tag{2.68}$$

Consequently, starting from (2.67) and noticing that $(k+1)(k+2) \leqslant 2(3)^k$ for any nonnegative integer k, we find that for any x in $[0,1]$,

$$h(-x) \geqslant \frac{x^2}{2} \sum_{k=0}^{\infty} \left(\frac{x}{3}\right)^k = \frac{x^2}{2(1-x/3)}. \tag{2.69}$$

In a similar way, $k+2 \leqslant 2(3/2)^k$ for any nonnegative integer k. Hence, it follows from (2.68) that for any positive x,

$$vh\left(\frac{x}{v}\right) \geqslant \frac{x^2}{2(v+x)} \sum_{k=0}^{\infty} \left(\frac{2x}{3(v+x)}\right)^k = \frac{x^2}{2(v+x/3)}. \tag{2.70}$$

Finally, putting together (2.69) and (2.70) into (2.66), we obtain that for any x in $[0,1]$,

$$(1+v)L_v^*(x) \geqslant \frac{x^2}{2}\left(\frac{1}{1-x/3}+\frac{1}{v+x/3}\right) = \frac{(1+v)x^2}{2(v+x/3)(1-x/3)},$$

which completes the proof of Lemma 2.32. $\qquad\qquad\square$

2.5 SubGaussian inequalities

This section is devoted to concentration inequalities with Gaussian rate functions. Let us recall Hoeffding's inequality for independent and bounded random variables, given in Section 2.2. Let $X_1,\ldots,X_n$ be a finite sequence of independent and centered random variables satisfying, for all $1 \leqslant k \leqslant n$, $a_k \leqslant X_k \leqslant b_k$ almost surely for some real constants a_k and b_k such that $a_k < b_k$. Let $S_n = X_1 + \cdots + X_n$. Then, for any positive x,

$$\mathbb{P}(S_n \geqslant x) \leqslant \exp\left(-\frac{2x^2}{D_n}\right) \qquad \text{where} \qquad D_n = \sum_{k=1}^{n}(b_k - a_k)^2. \qquad (2.71)$$

Our goal in this section is twofold. First, our aim is to weaken the assumptions on the random variables appearing in Hoeffding's inequality. Next, our aim is to obtain smaller constants than D_n in this inequality. This section is divided into four subsections. In Subsection 2.5.1, we are interested in Gaussian rate functions for the deviations on the right of sums of independent random variables bounded from above. In the next subsection, we apply the results of Subsection 2.5.1 and Section 2.3 to the deviation on the left of sums of independent and nonnegative random variables. In Subsection 2.5.3, we establish subGaussian inequalities for sums of random variables satisfying symmetric boundedness conditions. Finally, Subsection 2.5.4 is devoted to several improvements of Hoeffding's inequality (2.71). In particular, this subsection includes the so-called Kearns-Saul inequality [12] and the following improvement of Hoeffding's inequality. For any positive x,

$$\mathbb{P}(S_n \geqslant x) \leqslant \exp\left(-\frac{3x^2}{D_n + 2V_n}\right) \qquad (2.72)$$

where V_n stands for the variance of S_n. To the best of our knowledge, this improvement is new. In fact, this improvement will be derived from inequalities for sums of random variables bounded from above. This is the reason why we start this section by inequalities for sums of random variables bounded from above, which are the fundamental tools of Section 2.5.

2.5.1 Random variables bounded from above

Throughout this subsection, we assume that $X_1, \ldots, X_n$ is a finite sequence of independent and centered random variables satisfying, for all $1 \leqslant k \leqslant n$,

$$\operatorname{Var}(X_k) \leqslant v_k \qquad \text{and} \qquad X_k \leqslant b_k \tag{2.73}$$

almost surely, for some finite sequences $v_1, \ldots, v_n$ and $b_1, \ldots, b_n$ of positive real numbers. We start with our main result.

Theorem 2.33. *Let $X_1, \ldots, X_n$ be a finite sequence of independent and centered random variables satisfying (2.73). Denote $S_n = X_1 + \cdots + X_n$ and $V_n = v_1 + \cdots + v_n$. Let φ be the function defined by*

$$\varphi(v) = \begin{cases} \dfrac{1 - v^2}{|\log(v)|} & \text{if } v < 1, \\ 2v & \text{if } v \geqslant 1. \end{cases} \tag{2.74}$$

Then, for any positive x,

$$\mathbb{P}(S_n \geqslant x) \leqslant \exp\left(-\frac{x^2}{A_n}\right) \quad \text{where} \quad A_n = \sum_{k=1}^{n} b_k^2 \varphi\left(\frac{v_k}{b_k^2}\right). \tag{2.75}$$

Consequently, for any positive x,

$$\mathbb{P}(S_n \geqslant x) \leqslant \exp\left(-\frac{3x^2}{6V_n + B_n}\right) \leqslant \exp\left(-\frac{3x^2}{5V_n + C_n}\right) \tag{2.76}$$

where

$$B_n = \sum_{k=1}^{n} \left(b_k - \frac{v_k}{b_k}\right)_+^2 \quad \text{and} \quad C_n = \sum_{k=1}^{n} \max\left(b_k^2, v_k\right).$$

Remark 2.34. The function φ is continuous and increasing.

Remark 2.35. It immediately follows from (2.76) that, for any positive x,

$$\mathbb{P}(S_n \geqslant x) \leqslant \left(-\frac{x^2}{2C_n}\right). \tag{2.77}$$

This inequality was stated and proved in the more general framework of martingale difference sequences bounded from above, by Bentkus [5]. On the one hand, assume that for all $1 \leqslant k \leqslant n$, $v_k \geqslant b_k^2$. Then, the two above inequalities are equivalent to

$$\mathbb{P}(S_n \geqslant x) \leqslant \exp\left(-\frac{x^2}{2V_n}\right).$$

On the other hand, (2.76) is more efficient than (2.77) if $V_n < b_1^2 + \cdots + b_n^2$.

The proof of Theorem 2.33 relies on the two lemmas below, which will play an important role in the rest of the section. The first one gives an upper bound for the Laplace transform of a centered random variable bounded for above.

Lemma 2.36. *Let X be a centered random variable such that $X \leqslant 1$ almost surely and $\mathrm{Var}(X) \leqslant v$ for some positive constant v. Then, for any positive t,*

$$\log \mathbb{E}[\exp(tX)] \leqslant \frac{1}{4}\,\varphi(v)t^2 \tag{2.78}$$

where φ is the function given by (2.74).

Proof. According to Lemma 2.23, we have for any nonnegative t,

$$\log \mathbb{E}[\exp(tX)] \leqslant L_v(t)$$

where

$$L_v(t) = \log(ve^t + e^{-vt}) - \log(1+v).$$

Moreover, we already saw in Lemma 2.26 that the Legendre-Fenchel transform L_v^* of L_v satisfies $L_v^*(x) \geqslant \psi(x)$ for any positive x, where $\psi(x) = x^2/\varphi(v)$. Taking the Legendre-Fenchel transforms in this inequality, we then derive that $(L_v^*)^*(t) = L_v(t) \leqslant \psi^*(t)$ for any positive t. Moreover, it is not hard to see that, for any positive t, $\psi^*(t) = \varphi(v)t^2/4$, leading to

$$L_v(t) \leqslant \frac{1}{4}\,\varphi(v)t^2,$$

which completes the proof of Lemma 2.36. $\qquad\qquad\qquad\qquad\qquad\square$

The second lemma provides a computationally tractable upper bound for the function φ.

Lemma 2.37. *For any v in $]0,1]$,*

$$\varphi(v) \leqslant \frac{1}{3}(1+4v+v^2). \tag{2.79}$$

Proof. By the very definition (2.74) of the function φ, inequality (2.79) holds if and only if, for any v in $]0,1[$,

$$-(1+4v+v^2)\log v \geqslant 3(1-v^2).$$

Via the change of variables $u = 1 - v$, the above inequality is equivalent to

$$-(6-6u+u^2)\log(1-u) + 3u(u-2) \geqslant 0.$$

However, we deduce from the Taylor expansion of the logarithm

$$-\log(1-u) = \sum_{k=0}^{\infty} \frac{u^k}{k}$$

that, for any u in $]0,1[$,

$$-(6-6u+u^2)\log(1-u)+3u(u-2) = \sum_{k=5}^{\infty} \frac{(k-3)(k-4)}{k(k-1)(k-2)}\, u^k \geqslant 0,$$

which is exactly what we wanted to prove. $\qquad\square$

Proof of Theorem 2.33. We deduce from Lemma 2.36 together with the independence of the random variables $X_1,\ldots,X_n$ that, for any positive t,

$$\log\mathbb{E}\Big[\exp(tS_n)\Big] = \sum_{k=1}^{n}\log\mathbb{E}\big[\exp(tX_k)\big] = \sum_{k=1}^{n}\log\mathbb{E}\Big[\exp\Big(tb_k\frac{X_k}{b_k}\Big)\Big],$$

$$\leqslant \frac{t^2}{4}\sum_{k=1}^{n} b_k^2\varphi\Big(\frac{v_k}{b_k^2}\Big). \tag{2.80}$$

Consequently, (2.75) immediately follows (2.80) via the usual Chernoff calculation. It only remains to prove (2.76). According to Lemma 2.37, we have for any $1 \leqslant k \leqslant n$, as soon as $v_k < b_k^2$,

$$3b_k^2\varphi\Big(\frac{v_k}{b_k^2}\Big) \leqslant b_k^2 + 4v_k + \frac{v_k^2}{b_k^2} = 6v_k + \Big(b_k - \frac{v_k}{b_k}\Big)^2.$$

Hence, for any $1 \leqslant k \leqslant n$ such that $v_k < b_k^2$,

$$3b_k^2\varphi\Big(\frac{v_k}{b_k^2}\Big) \leqslant \Big(6v_k + \Big(b_k - \frac{v_k}{b_k}\Big)_+^2\Big).$$

Obviously, the above inequality still holds if $v_k \geqslant b_k^2$, as the left-hand side in this inequality is exactly $6v_k$. Therefore,

$$A_n = \sum_{k=1}^{n} b_k^2\varphi\Big(\frac{v_k}{b_k^2}\Big) \leqslant \frac{1}{3}\sum_{k=1}^{n} 6v_k + \Big(b_k - \frac{v_k}{b_k}\Big)_+^2 = \frac{1}{3}(6V_n + B_n),$$

which leads to the first inequality in (2.76). Finally, for any $1 \leqslant k \leqslant n$, as

$$\Big(b_k - \frac{v_k}{b_k}\Big)_+^2 \leqslant b_k\Big(b_k - \frac{v_k}{b_k}\Big)_+ = (b_k^2 - v_k)_+,$$

we find that

$$6V_n + B_n \leqslant 5V_n + \sum_{k=1}^{n} v_k + (b_k^2 - v_k)_+ = 5V_n + \sum_{k=1}^{n} \max(b_k^2, v_k) = 5V_n + C_n$$

which clearly implies the second inequality in (2.76). $\qquad\square$

We shall now apply Theorem 2.33 to weighted sums of independent random variables bounded from above. It yields an exponential inequality with a Gaussian rate function.

Corollary 2.38. *Let $Z_1, \ldots, Z_n$ be a finite sequence of independent and centered random variables such that, for all $1 \leqslant k \leqslant n$, $Z_k \leqslant 1$ almost surely. Assume that there exists some positive real v such that, for all $1 \leqslant k \leqslant n$, $\mathbb{E}[Z_k^2] \leqslant v$. Denote $S_n = b_1 Z_1 + \cdots + b_n Z_n$ for some positive real numbers $b_1, \ldots, b_n$ and let $\|b\|_2 = (b_1^2 + \cdots + b_n^2)^{1/2}$. Then, for any positive x,*

$$\mathbb{P}(S_n \geqslant \|b\|_2 x) \leqslant \exp(-g(v)x^2) \tag{2.81}$$

where

$$g(v) = \begin{cases} \dfrac{|\log(v)|}{1 - v^2} & \text{if } v < 1, \\ \dfrac{1}{2v} & \text{if } v \geqslant 1. \end{cases}$$

Remark 2.39. One can observe that for any positive v, $g(v) = 1/\varphi(v)$. Hence, the function g is decreasing. In particular, if $v \leqslant 1$, then $g(v) \geqslant 1/2$.

Remark 2.40. For $v \geqslant 1$, we will give a more efficient inequality in Section 2.6.

Proof. Corollary 2.38 immediately follows from inequality (2.75) applied to the sequence $X_1, \ldots, X_n$ where, for all $1 \leqslant k \leqslant n$, $X_k = b_k Z_k$ with $v_k = b_k^2 v$. $\qquad \square$

2.5.2 Nonnegative random variables

We shall now focus our attention on concentration inequalities for nonnegative random variables. Throughout this subsection, we assume that $X_1, \ldots, X_n$ is a finite sequence of nonnegative independent random variables with finite variances. For all $1 \leqslant k \leqslant n$, let

$$m_k = \mathbb{E}[X_k] \qquad \text{and} \qquad v_k = \text{Var}(X_k).$$

Theorem 2.33 leads to the result below for the deviation on the left of sums of nonnegative random variables.

Theorem 2.41. *Let $X_1, \ldots, X_n$ be a finite sequence of independent and nonnegative random variables with finite variances. Denote $S_n = X_1 + \cdots + X_n$ and $V_n = \text{Var}(S_n)$. Then, for any positive x,*

$$\mathbb{P}(S_n \leqslant \mathbb{E}[S_n] - x) \leqslant \exp\left(-\frac{x^2}{2V_n + W_n}\right) \tag{2.82}$$

where

$$W_n = \frac{1}{3} \sum_{k=1}^{n} \left(\frac{m_k^2 - v_k}{m_k} \right)_+^2. \tag{2.83}$$

Proof. For all $1 \leqslant k \leqslant n$, let Z_k be the random variable defined by $Z_k = m_k - X_k$. The sequence $Z_1, \ldots, Z_n$ satisfies the assumptions of Theorem 2.33 with $b_k = \mathbb{E}[X_k]$ and $v_k = \mathrm{Var}(X_k)$. Hence, (2.82) clearly follows from the first inequality in (2.76). $\qquad\square$

Remark 2.42. Let $\mathcal{V}_n = \mathbb{E}[X_1^2] + \cdots + \mathbb{E}[X_n^2]$. It is not hard to see that $V_n + 3W_n \leqslant \mathcal{V}_n$. Hence, $2V_n + W_n \leqslant (5V_n + \mathcal{V}_n)/3 \leqslant 2\mathcal{V}_n$. Consequently, one can replace $2V_n + W_n$ by $2\mathcal{V}_n$ in (2.82). The inequality with the denominator $2\mathcal{V}_n$ may be found in Maurer [16].

Example 2.43. Let $X_1, \ldots, X_n$ be a finite sequence of independent random variables sharing the same Exponential $\mathscr{E}(\lambda)$ distribution with $\lambda > 0$. In that case, $W_n = 0$, which ensures that for any positive x,

$$\mathbb{P}(S_n \leqslant \mathbb{E}[S_n] - x) \leqslant \exp\left(-\frac{x^2}{2V_n} \right).$$

In Section 2.7, we will give more efficient inequalities for sums of independent random variables with exponential distributions. This is the reason why we give a second example below.

Example 2.44. Let $\varepsilon_1, \ldots, \varepsilon_n$ be a finite sequence of independent Poisson random variables and let $B_1, \ldots, B_n$ be a finite sequence of independent Bernoulli random variables. Assume that these two sequences are mutually independent. In addition, suppose that, for each $1 \leqslant k \leqslant n$, ε_k has the Poisson $\mathscr{P}(\lambda_k)$ distribution, while B_k has the Bernoulli $\mathscr{B}(p_k)$ distribution with $\lambda_k = k+1$ and $p_k = k/(k+1)$. For all $1 \leqslant k \leqslant n$, let X_k be the random variable defined by $X_k = B_k \varepsilon_k$. One can easily check that, for all $1 \leqslant k \leqslant n$, $\mathbb{E}[X_k] = k$, $\mathbb{E}[X_k^2] = k(k+2)$, which implies that $v_k = 2k$ and $(m_k^2 - v_k)_+ = k \max(0, k-2)$. Hence, for any $n \geqslant 2$, $V_n = n(n+1)$,

$$W_n = \frac{1}{18}(n-1)(n-2)(2n-3) \qquad \text{and} \qquad \mathcal{V}_n = \frac{1}{6}n(n+1)(2n+7).$$

According to Theorem 2.41, we find that for any positive x,

$$\mathbb{P}(S_n \leqslant \mathbb{E}[S_n] - x) \leqslant \exp\left(-\frac{18x^2}{2n^3 + 27n^2 + 49n - 6} \right)$$

Under the same assumptions, Maurer's inequality yields

$$\mathbb{P}(S_n \leqslant \mathbb{E}[S_n] - x) \leqslant \exp\left(-\frac{3x^2}{n(2n^2 + 9n + 7)} \right).$$

When $n = 10$, the first upper bound is equal to $\exp(-x^2/288)$, while the second upper bound is equal to $\exp(-x^2/990)$. For example, if $x = 40$, the first bound is equal to 3.866×10^{-3}, while the second bound is equal to 1.986×10^{-1}.

2.5.3 Symmetric conditions for bounded random variables

Throughout this subsection, we assume that $X_1, \ldots, X_n$ is a finite sequence of independent and centered random variables satisfying, for all $1 \leqslant k \leqslant n$,

$$|X_k| \leqslant b_k \tag{2.84}$$

almost surely, for some finite sequence $b_1, \ldots, b_n$ of positive real numbers. As usual, denote $S_n = X_1 + \cdots + X_n$. In order to state our concentration inequality for S_n, we need the elementary observation: it follows from condition (2.84) that, for all $1 \leqslant k \leqslant n$,

$$v_k = \mathrm{Var}(X_k) \leqslant b_k^2. \tag{2.85}$$

The result below is an immediate consequence of Theorem 2.33. The proof being obvious is omitted.

Theorem 2.45. *Let $X_1, \ldots, X_n$ be a finite sequence of independent and centered random variables satisfying (2.84). Denote $S_n = X_1 + \cdots + X_n$ and $V_n = \mathrm{Var}(S_n)$. Let φ be the function defined, for v in $]0, 1[$, by*

$$\varphi(v) = \frac{1 - v^2}{|\log(v)|} \tag{2.86}$$

with $\varphi(1) = 2$. Then, for any positive x,

$$\mathbb{P}(S_n \geqslant x) \leqslant \exp\left(-\frac{x^2}{A_n}\right) \leqslant \exp\left(-\frac{3x^2}{5V_n + B_n}\right) \leqslant \exp\left(-\frac{x^2}{2B_n}\right) \tag{2.87}$$

where

$$A_n = \sum_{k=1}^{n} b_k^2 \varphi\left(\frac{v_k}{b_k^2}\right) \qquad and \qquad B_n = \sum_{k=1}^{n} b_k^2.$$

Remark 2.46. Note that for any v in $]0, 1[$, $\varphi(v) < 2$. Consequently, $A_n < 2B_n$ unless all the random variables $X_1, \ldots, X_n$ have the Bernoulli distribution given, for all $1 \leqslant k \leqslant n$, by $\mathbb{P}(X_k = -b_k) = \mathbb{P}(X_k = b_k) = 1/2$.

2.5.4 Asymmetric conditions for bounded random variables

This subsection is devoted to improvements of Hoeffding's inequality. We start by an improvement of Hoeffding's inequality which holds for non-centered random variables. Up to our knowledge, this result is new.

Theorem 2.47. *Let $X_1, \ldots, X_n$ be a finite sequence of independent random variables. Assume that for all $1 \leqslant k \leqslant n$, one can find two constants $a_k < b_k$ such that $a_k \leqslant X_k \leqslant b_k$ almost surely. Denote $S_n = X_1 + \cdots + X_n$ and $V_n = Var(S_n)$. Then, for any positive x,*

$$\mathbb{P}(S_n - \mathbb{E}[S_n] \geqslant x) \leqslant \exp\left(-\frac{3x^2}{D_n + 2V_n}\right) \tag{2.88}$$

where

$$D_n = \sum_{k=1}^{n} (b_k - a_k)^2.$$

Remark 2.48. As mentioned in Remark 2.17, $D_n \geqslant 4V_n$. Consequently, Theorem 2.47 clearly improves Theorem 2.16. One can observe that $D_n > 4V_n$ except if all the random variables $X_1, \ldots, X_n$ have the Bernoulli distribution given, for all $1 \leqslant k \leqslant n$, by $\mathbb{P}(X_k = a_k) = \mathbb{P}(X_k = b_k) = 1/2$.

Proof. For all $1 \leqslant k \leqslant n$, let Y_k be the random variable defined by $Y_k = X_k - \mathbb{E}[X_k]$. One can observe that $Y_1, \ldots, Y_n$ is a sequence of independent and centered random variables such that, for all $1 \leqslant k \leqslant n$, $\alpha_k \leqslant Y_k \leqslant \beta_k$ almost surely with $\alpha_k = a_k - \mathbb{E}[X_k]$ and $\beta_k = b_k - \mathbb{E}[X_k]$. In addition, for all $1 \leqslant k \leqslant n$, $\alpha_k < 0$ and $\beta_k > 0$, except if $Y_k = 0$ almost surely, which means that Y_k can be removed. For all $1 \leqslant k \leqslant n$, we have $|Y_k| \leqslant c_k$ almost surely where $c_k = \max(-\alpha_k, \beta_k)$, and $v_k = Var(X_k) = Var(Y_k)$. Consequently, the sequence $Y_1, \ldots, Y_n$ satisfies the assumptions of Theorem 2.45 which ensures that, for any positive x,

$$\mathbb{P}(S_n - \mathbb{E}[S_n] \geqslant x) \leqslant \exp\left(-\frac{x^2}{A_n}\right) \quad \text{where} \quad A_n = \sum_{k=1}^{n} c_k^2 \varphi\left(\frac{v_k}{c_k^2}\right) \tag{2.89}$$

Furthermore, we deduce from Lemma 2.37 that

$$A_n \leqslant \frac{1}{3} \sum_{k=1}^{n} \left(c_k^2 + 4v_k + \frac{v_k^2}{c_k^2}\right) = \frac{1}{3} \sum_{k=1}^{n} \left(2v_k + \left(c_k + \frac{v_k}{c_k}\right)^2\right). \tag{2.90}$$

In order to prove (2.88), it only remains to show that $\Delta_n \leqslant D_n$ where

$$\Delta_n = \sum_{k=1}^{n} \left(c_k + \frac{v_k}{c_k}\right)^2.$$

According to Lemma 2.18, $Var(Y_k) \leqslant -\alpha_k \beta_k = \min(-\alpha_k, \beta_k) \max(-\alpha_k, \beta_k)$, which implies that

$$\Delta_n \leqslant \sum_{k=1}^{n} \left(\max(-\alpha_k, \beta_k) + \min(-\alpha_k, \beta_k)\right)^2 = \sum_{k=1}^{n} (\beta_k - \alpha_k)^2 = D_n. \tag{2.91}$$

Finally, (2.88) follows from (2.89), (2.90) and (2.91), which achieves the proof of Theorem 2.47.   □

To conclude this subsection, we give an improvement of Hoeffding's inequality due to Kearns and Saul [12]. This improvement takes into account the centering of the random variables $X_1, \ldots, X_n$.

Theorem 2.49. *Let $X_1, \ldots, X_n$ be a finite sequence of independent random variables and let $S_n = X_1 + \cdots + X_n$. Assume that for all $1 \leqslant k \leqslant n$, one can find two constants $a_k < b_k$ such that $a_k \leqslant X_k \leqslant b_k$ almost surely. For all $1 \leqslant k \leqslant n$, denote $m_k = \mathbb{E}[X_k]$,*

$$p_k = (b_k - m_k)^2 \varphi\left(\frac{m_k - a_k}{b_k - m_k}\right) \quad and \quad q_k = \frac{(b_k - a_k)(b_k + a_k - 2m_k)}{\log((b_k - m_k)/(m_k - a_k))}$$

with the convention that $q_k = (b_k - a_k)^2/2$ if $m_k = (a_k + b_k)/2$, where φ is the function given by (2.74). Then, for any positive x,

$$\mathbb{P}(S_n - \mathbb{E}[S_n] \geqslant x) \leqslant \exp\left(-\frac{x^2}{P_n}\right) \leqslant \exp\left(-\frac{x^2}{Q_n}\right) \tag{2.92}$$

where

$$P_n = \sum_{k=1}^{n} p_k \quad and \quad Q_n = \sum_{k=1}^{n} q_k.$$

Remark 2.50. One can observe that, for all $1 \leqslant k \leqslant n$, $q_k < (b_k - a_k)^2/2$ as soon as $m_k \neq (a_k + b_k)/2$. Note also that the coefficients q_k are symmetric functions of (a_k, b_k). Hence, we obtain that, for any positive x,

$$\mathbb{P}(|S_n - \mathbb{E}[S_n]| \geqslant x) \leqslant 2\exp\left(-\frac{x^2}{Q_n}\right).$$

Proof. We shall proceed as in the proof of Theorem 2.47. We already saw that, for all $1 \leqslant k \leqslant n$, $Y_k \leqslant \beta_k$ almost surely and $\mathrm{Var}(Y_k) \leqslant -\alpha_k \beta_k$ with $\alpha_k = a_k - \mathbb{E}[X_k]$ and $\beta_k = b_k - \mathbb{E}[X_k]$. Then, we immediately deduce from inequality (2.75) that, for any positive x,

$$\mathbb{P}(S_n - \mathbb{E}[S_n] \geqslant x) \leqslant \exp\left(-\frac{x^2}{P_n}\right).$$

It only remains to prove that $P_n \leqslant Q_n$. If $-\alpha_k \leqslant \beta_k$, then $p_k = q_k$ and there is nothing to prove. On the other side, if $-\alpha_k > \beta_k$, we may apply the elementary inequality $\log(x) \leqslant (x - 1/x)/2$, which holds if $x > 1$, to $x = -\alpha_k/\beta_k$. This inequality ensures that

$$\log\left(-\frac{\alpha_k}{\beta_k}\right) \leqslant \frac{\beta_k^2 - \alpha_k^2}{2\alpha_k \beta_k},$$

leading to $p_k = -2\alpha_k\beta_k \leqslant q_k$. Finally, we have shown that $P_n \leqslant Q_n$, which completes the proof of Theorem 2.49 $\qquad\qquad\square$

2.6 Always a little further on weighted sums

The goal of this section is to go a little bit further on concentration inequalities for weighted sums. Let $Z_1,\ldots,Z_n$ be a finite sequence of independent and centered random variables with finite Laplace transform on a right neighborhood of the origin. More precisely, we will assume that there exists a convex and increasing function ℓ from $[0,+\infty[$ to $[0,\infty]$ such that $\ell(0) = \ell'(0) = 0$ and, for all $1 \leqslant k \leqslant n$, and for any $t \geqslant 0$,

$$\log\mathbb{E}\big[\exp(tZ_k)\big] \leqslant \ell(t). \tag{2.93}$$

Our aim is to establish one-sided deviation inequalities for $S_n = b_1 Z_1 + \cdots + b_n Z_n$, for some positive real numbers $b_1,\ldots,b_n$. As usual, we denote for all $p \geqslant 1$,

$$\|b\|_p = \big(b_1^p + b_2^p + \cdots + b_n^p\big)^{1/p}$$

and $\|b\|_\infty = \max(b_1,b_2,\ldots,b_n)$. One can notice that there are only a few results on that direction. In a paper devoted to McDiarmid's inequality, Rio [20] uses the concavity of ℓ' to establish an upper bound for $\mathbb{P}(S_n \geqslant x)$. Below, we give a more general version of this inequality and another inequality for functions ℓ with a convex derivative.

Theorem 2.51. *Let $Z_1,\ldots,Z_n$ be a finite sequence of independent and centered random variables such that, for all $1 \leqslant k \leqslant n$, the random variable Z_k satisfies (2.93). Denote $S_n = b_1 Z_1 + \cdots + b_n Z_n$ for some positive real numbers $b_1,\ldots,b_n$.*

1) If the function ℓ has a concave derivative, then, for any positive x,

$$\mathbb{P}(S_n \geqslant x) \leqslant \exp\Big(-\frac{\|b\|_1^2}{\|b\|_2^2}\,\ell^*\Big(\frac{x}{\|b\|_1}\Big)\Big), \tag{2.94}$$

where ℓ^* stands for the Legendre-Fenchel dual of ℓ.

2) If the function h defined, for any positive t, by $h(t) = \ell'(t)/t$, is nondecreasing on $\mathbb{R}^+$, then, for any positive x,

$$\mathbb{P}\Big(S_n \geqslant x\Big) \leqslant \exp\Big(-\frac{\|b\|_2^2}{\|b\|_\infty^2}\,\ell^*\Big(\frac{\|b\|_\infty x}{\|b\|_2^2}\Big)\Big). \tag{2.95}$$

Remark 2.52. If the function ℓ has a convex derivative, then h is nondecreasing, and consequently (2.95) holds true.

Proof. We clearly obtain from (2.93) that, for any positive t,

$$\log\mathbb{E}\big[\exp(tS_n)\big] \leqslant \ell(b_1 t) + \cdots + \ell(b_n t). \tag{2.96}$$

Starting from (2.96), we now prove (2.95). As the function h is nondecreasing, we have for any $1 \leqslant k \leqslant n$,

$$\ell(b_k t) = \int_0^t b_k \ell'(b_k x) dx \leqslant \int_0^t \frac{b_k^2}{\|b\|_\infty} \ell'(\|b\|_\infty x) dx = \frac{b_k^2}{\|b\|_\infty^2} \ell(\|b\|_\infty t),$$

which ensures that, for any positive t,

$$\ell(b_1 t) + \cdots + \ell(b_n t) \leqslant \frac{\|b\|_2^2}{\|b\|_\infty^2} \ell(\|b\|_\infty t). \tag{2.97}$$

It follows from Markov's inequality together with (2.96) and (2.97) that, for any positive t,

$$\log \mathbb{P}(S_n \geqslant x) \leqslant -xt + \frac{\|b\|_2^2}{\|b\|_\infty^2} \ell(\|b\|_\infty t) = -\frac{\|b\|_2^2}{\|b\|_\infty^2} \left(\frac{\|b\|_\infty xs}{\|b\|_2^2} - \ell(s) \right)$$

where $s = \|b\|_\infty t$. Hence, we immediately deduce (2.95) taking the infimum over all positive reals s. We now proceed to the proof of (2.94). Since $\ell(0) = 0$, we have for any positive x,

$$\ell(b_1 x) + \cdots + \ell(b_n x) = \int_0^x \big(b_1 \ell'(b_1 t) + \cdots + b_n \ell'(b_n t) \big) dt.$$

Next, the concavity of ℓ' implies that

$$b_1 \ell'(b_1 t) + \cdots + b_n \ell'(b_n t) \leqslant \|b\|_1 \ell' \left(\frac{\|b\|_2^2 t}{\|b\|_1} \right).$$

Hence, for any positive t,

$$\ell(b_1 t) + \cdots + \ell(b_n t) \leqslant \frac{\|b\|_1^2}{\|b\|_2^2} \ell \left(\frac{\|b\|_2^2 t}{\|b\|_1} \right). \tag{2.98}$$

Therefore, we infer from Markov's inequality together with (2.96) and (2.98) that, for any positive t,

$$\log \mathbb{P}(S_n \geqslant x) \leqslant -xt + \frac{\|b\|_1^2}{\|b\|_2^2} \ell \left(\frac{\|b\|_2^2 t}{\|b\|_1} \right) = -\frac{\|b\|_1^2}{\|b\|_2^2} \left(\frac{xs}{\|b\|_1} - \ell(s) \right)$$

where $s = (\|b\|_2^2 / \|b\|_1) t$. Finally, we obtain (2.94) taking once again the infimum over all positive reals s. $\qquad\square$

Example 2.53. Let $Z_1, \ldots, Z_n$ be a finite sequence of independent and centered random variables such that, for all $1 \leqslant k \leqslant n$, $Z_k \leqslant 1$ almost surely. Assume that there exists some real number $v \geqslant 1$ such that, for all $1 \leqslant k \leqslant n$, $\mathbb{E}[Z_k^2] \leqslant v$. Then, it follows from Lemma 2.23 that, for all $1 \leqslant k \leqslant n$ and for any positive t,

$$\log \mathbb{E}[\exp(tZ_k)] \leqslant L_v(t) \quad \text{where} \quad L_v(t) = \log(ve^t + e^{-vt}) - \log(1+v).$$

For any positive t, denote $f(t) = \log(ve^t + 1) - \log(1+v)$ with this notation, we have $L_v(t) = f((1+v)t) - vt$. Consequently, the function L_v' is concave on $\mathbb{R}^+$ if and only if f' is concave on $\mathbb{R}^+$. However, it is not hard to see that for any positive t,

$$f^{(3)}(t) = ve^t(1 - ve^t)(1 + ve^t)^{-3} < 0.$$

Hence, f' and L_v' are concave. Hence, we deduce from (2.55) and (2.94) that, for any x in $[0,1]$,

$$\mathbb{P}(S_n \geqslant \|b\|_1 x) \leqslant \exp\left(-\frac{\|b\|_1^2}{\|b\|_2^2}\left(\frac{v+x}{v+1}\log\left(1+\frac{x}{v}\right) + \frac{1-x}{1+v}\log(1-x)\right)\right). \quad (2.99)$$

For example, assume that the random variables $Z_1,\ldots,Z_n$ are with values in $[-v,1]$ for some $v \geqslant 1$. Then, it follows from Lemma 2.18 that, for all $1 \leqslant k \leqslant n$, $\mathbb{E}[Z_k^2] \leqslant v$, which means that the above inequality holds true.

Example 2.54. Let $Z_1,\ldots,Z_n$ be a finite sequence of independent and centered random variables satisfying a Bernstein's type condition: there exists some real v in $]0,1]$ such that, for any $1 \leqslant k \leqslant n$, $\mathbb{E}[Z_k^2] \leqslant v$ and, for any integer $p \geqslant 3$,

$$\mathbb{E}[(\max(0,Z_k))^p] \leqslant \frac{p!v}{2}.$$

According to (2.14) with $c = 1$, we have for all $1 \leqslant k \leqslant n$ and for any t in $[0,1[$,

$$\log \mathbb{E}[\exp(tZ_k)] \leqslant \ell_v(t) \quad \text{where} \quad \ell_v(t) = \log\left(1 + \frac{vt^2}{2(1-t)}\right).$$

Hereafter, let h_v be the function defined, for any positive t, by $h_v(t) = \ell_v'(t)/t$. It is not hard to see that

$$\log(h_v(t)) = \log(v/2) + \log(2-t) - \log(1-t) - \log(1-t+vt^2/2)$$

which leads to

$$\log(h_v(t))' = \frac{1}{(2-t)(1-t)} + \frac{1-vt}{1-t+vt^2/2} > 0$$

for any t in $[0,1[$, since $vt \leqslant 1$ for all t in $[0,1]$. Consequently, we obtain from (2.15) and (2.95) in the particular case $\|b\|_\infty = 1$ that, for any positive x,

$$\mathbb{P}(S_n \geqslant \|b\|_2^2 x) \leqslant \exp\left(-\|b\|_2^2\left(\frac{x^2}{v+x} - \log\left(1 + \frac{x^2}{2(v+x)}\right)\right)\right).$$

This inequality cannot be deduced from Theorem 2.1.

2.7 Sums of Gamma random variables

In this section, we are interested in deviation inequalities for sums of independent nonnegative random variables with Gamma distributions. Let a and b be some positive real numbers. A random variable X has the $\Gamma(a,b)$ distribution if its probability density function f_X is given by

$$f_X(x) = \frac{x^{a-1}\exp(-x/b)}{b^a \Gamma(a)} \mathrm{I}_{\{x>0\}} \tag{2.100}$$

where

$$\Gamma(a) = \int_0^\infty x^{a-1}\exp(-x)\,dx$$

stands for the Euler's Gamma function. From the definition, if X has the $\Gamma(a,b)$ distribution, then X/b has the $\Gamma(a,1)$ distribution. One can observe that the $\Gamma(1,b)$ distribution coincides with the Exponential $\mathscr{E}(1/b)$ distribution. Moreover, if Z is distributed as a normal $\mathscr{N}(0,1)$ random variable, then Z^2 has the $\Gamma(1/2,2)$ distribution. Consequently, sums of Gamma random variables include sums of exponential random variables as well as sums of weighted chi-square random variables.

As shown by the lemma below, the Gamma random variables have a nice Laplace transform.

Lemma 2.55. *Let X be a random variable with $\Gamma(a,b)$ distribution. Then, for any real t,*

$$\log \mathbb{E}[\exp(tX)] = a\ell(bt) \tag{2.101}$$

where ℓ is the strictly convex function given by

$$\ell(t) = \begin{cases} -\log(1-t) & \text{if } t < 1, \\ +\infty & \text{if } t \geqslant 1. \end{cases}$$

Remark 2.56. We immediately deduce from Lemma 2.55 that $\mathbb{E}[X] = ab\ell'(0) = ab$ and $\mathrm{Var}(X) = ab^2\ell''(0) = ab^2$.

Proof. It is enough to prove Lemma 2.55 in the special case $b = 1$. It follows from (2.100) that for any real t,

$$\mathbb{E}[\exp(tX)] = \frac{1}{\Gamma(a)} \int_0^\infty x^{a-1}\exp(-x(1-t))\,dx.$$

From the above equality, $\mathbb{E}[\exp(tX)] = \infty$ if $t \geqslant 1$. Moreover, if $t < 1$, we obtain via the change of variables $y = x(1-t)$ that

$$\mathbb{E}[\exp(tX)] = \frac{(1-t)^{-a}}{\Gamma(a)} \int_0^\infty y^{a-1}\exp(-y)\,dy = (1-t)^{-a},$$

which proves Lemma 2.55. $\square$

Hereafter, let $X_1, \ldots X_n$ be a finite sequence of independent random variables such that, for all $1 \leqslant k \leqslant n$, X_k has the $\Gamma(a_k, b_k)$ distribution, where a_k and b_k are positive real numbers. Define

$$\|b\|_{1,a} = \sum_{k=1}^{n} a_k b_k \quad \text{and} \quad \|b\|_{2,a} = \Big(\sum_{k=1}^{n} a_k b_k^2 \Big)^{1/2}. \tag{2.102}$$

We now state our concentration inequalities for sums of Gamma random variables.

Theorem 2.57. *Let $X_1, \ldots X_n$ be a finite sequence of independent random variables such that, for all $1 \leqslant k \leqslant n$, X_k has the $\Gamma(a_k, b_k)$ distribution. Let $S_n = X_1 + \cdots + X_n$. Then, for any positive x,*

$$\mathbb{P}(S_n \geqslant \|b\|_{1,a} + x \|b\|_{2,a}^2) \leqslant \exp\Big(-\frac{\|b\|_{2,a}^2}{\|b\|_\infty^2} \big(\|b\|_\infty x - \log(1 + \|b\|_\infty x) \big) \Big) \tag{2.103}$$

$$\leqslant \exp\Big(-\frac{\|b\|_{2,a}^2 x^2}{1 + \|b\|_\infty x + \sqrt{1 + 2\|b\|_\infty x}} \Big). \tag{2.104}$$

In addition, for any x in $]0, 1[$,

$$\mathbb{P}(S_n \leqslant \|b\|_{1,a} - x \|b\|_{1,a}) \leqslant \exp\Big(\frac{\|b\|_{1,a}^2}{\|b\|_{2,a}^2} \big(\log(1-x) + x \big) \Big), \tag{2.105}$$

$$\leqslant \exp\Big(-\frac{\|b\|_{1,a}^2}{\|b\|_{2,a}^2} \frac{x^2}{2} \Big). \tag{2.106}$$

Remark 2.58. Note from Remark 2.56 that $\mathbb{E}[S_n] = \|b\|_{1,a}$ and $\mathrm{Var}(S_n) = \|b\|_{2,a}^2$. Consequently, Theorem 2.57 provides an exponential concentration inequality for S_n around its mean with the adequate variance term. Moreover, (2.104) is a Bernstein's type inequality, which may be found in the monograph of Boucheron, Lugosi, and Massart [7]. We now compare inequalities (2.103) and (2.104), as well as inequalities (2.105) and (2.106) in the particular case where the random variables $X_1, \ldots X_n$ share the same $\Gamma(a, 1)$ distribution for some positive a. Figure 2.6 compares the four rate functions

$$\Phi_R(x) = x - \log(1 + x) \quad \text{and} \quad \Psi_R(x) = \frac{x^2}{1 + x + \sqrt{1 + 2x}},$$

$$\Phi_L(x) = -x - \log(1 - x) \quad \text{and} \quad \Psi_L(x) = \frac{x^2}{2}$$

for x in the interval $]0, 4/5]$ while Figure 2.7 compares the rate functions Φ_L and Ψ_L for x in the interval $[4/5, 1[$. One can see in Figure 2.6 that (2.103) and (2.105) outperform (2.104) and (2.106), respectively. It is even more spectacular in Figure 2.7 that (2.105) is much more accurate than (2.106).

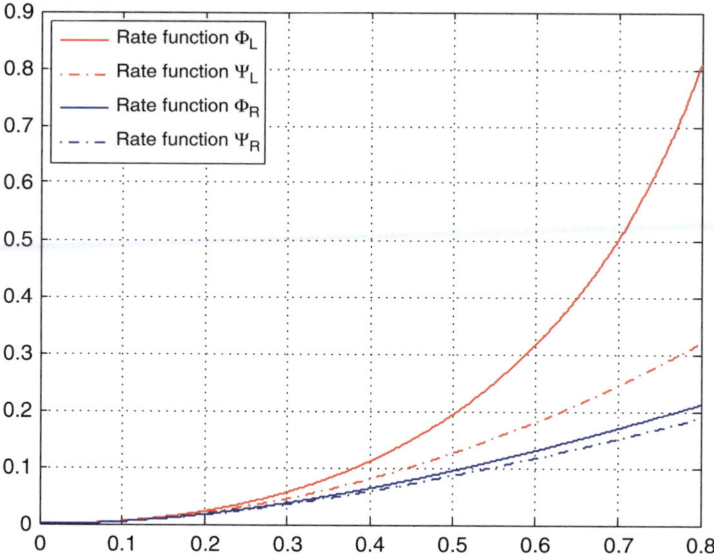

Fig. 2.6 Comparisons in Gamma concentration inequalities

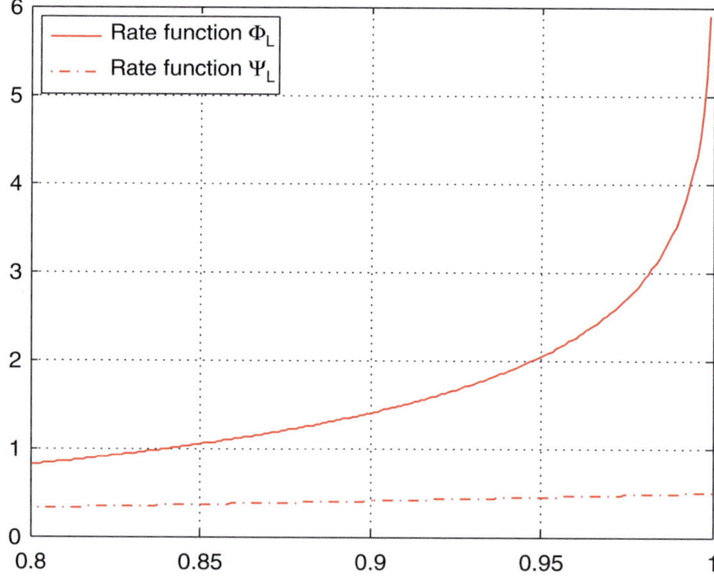

Fig. 2.7 Comparisons in Gamma concentration inequalities on the left side

Remark 2.59. Assume that $\|b\|_\infty = 1$. Starting from inequality (2.103) and using the upper bound

$$\Phi_R^{-1}(x) \leqslant \Lambda(x) = x + \log(1 + x + \sqrt{2x}) + \frac{\log(1 + x + \sqrt{2x}) - \sqrt{2x}}{x + \sqrt{2x}} \qquad (2.107)$$

given in Del Moral and Rio [9], one can prove that, for any positive x,

$$\mathbb{P}\big(S_n \geqslant \|b\|_{1,a} + \Lambda(x)\|b\|_{2,a}^2\big) \leqslant \exp\big(-\|b\|_{2,a}^2 x\big). \qquad (2.108)$$

Since $\Lambda(x) < x + \sqrt{2x}$, (2.108) is more efficient than the reversed form of (2.104).

Proof. We shall prove (2.103) in the particular case $\|b\|_\infty = 1$, inasmuch as the general case follows by dividing the initial random variables by $\|b\|_\infty$. Let ℓ_c be the convex function defined, for any real t, by $\ell_c(t) = \ell(t) - t$. We infer from Lemma 2.55 that, for any real t,

$$\log \mathbb{E}[\exp(t(S_n - \mathbb{E}[S_n]))] \leqslant \sum_{k=1}^{n} a_k \ell_c(b_k t). \qquad (2.109)$$

Moreover, let h_c be the function defined, for any positive t, by $h_c(t) = \ell_c(t)/t^2$. The function h_c is increasing on $]0, +\infty[$. Hence, for any positive t, $\ell_c(b_k t) \leqslant b_k^2 \ell_c(t)$, which implies that

$$\log \mathbb{E}[\exp(t(S_n - \mathbb{E}[S_n]))] \leqslant \|b\|_{2,a}^2 \ell_c(t). \qquad (2.110)$$

We deduce from Markov's inequality that for any positive x and for any t in $]0,1[$,

$$\log \mathbb{P}(S_n - \mathbb{E}[S_n] \geqslant x\|b\|_{2,a}^2) \leqslant -\|b\|_{2,a}^2(xt - \ell_c(t)). \qquad (2.111)$$

The optimal value t in the above inequality is given by the elementary equation $\ell_c'(t) = t/(1-t) = x$, leading to $t = x/(1+x)$. By taking this value of t, we find that

$$\ell_c^*(x) = xt - \ell_c(t) = (x+1)t + \log(1-t) = x - \log(1+x), \qquad (2.112)$$

which, via (2.111), achieves the proof of (2.103). One can observe from (2.112) that, for any $x \geqslant 0$, $\ell_c^*(x) = \ell_c(-x)$. From the reflexivity properties of the Legendre-Fenchel dual, this equality also holds true for any $x \leqslant 0$. The proof of (2.104) immediately follows from the fact that, for any positive t, $\ell_c(t) < \varphi(t) = t^2/(2 - 2t)$, which implies that for any positive x,

$$\ell_c^*(x) > \varphi^*(x) = \frac{x^2}{1 + x + \sqrt{1 + 2x}}.$$

We are in position to prove (2.105). It follows from (2.109) that, for any positive t,

$$\log \mathbb{E}[\exp(t(\mathbb{E}[S_n] - S_n))] \leqslant L(t) \qquad (2.113)$$

where

$$L(t) = \sum_{k=1}^{n} a_k \ell_c(-b_k t) = \sum_{k=1}^{n} a_k \ell_c^*(b_k t).$$

According to (2.112), $(\ell_c^*)'$ is a concave function as $(\ell_c^*)'(t) = t/(1+t)$. It ensures that, for any positive t,

$$L'(t) = \sum_{k=1}^{n} a_k b_k \left(\frac{b_k t}{1 + b_k t} \right) \leqslant \|b\|_{1,a} \left(\frac{\|b\|_{2,a}^2 t}{\|b\|_{1,a} + \|b\|_{2,a}^2 t} \right).$$

Integrating this inequality, we obtain that, for any positive t,

$$L(t) \leqslant \frac{\|b\|_{1,a}^2}{\|b\|_{2,a}^2} \ell_c^* \left(\frac{\|b\|_{2,a}^2 t}{\|b\|_{1,a}} \right). \tag{2.114}$$

Therefore, we find from (2.113) and (2.114) together with Markov's inequality that for any positive x and for any positive t,

$$\log \mathbb{P}(\mathbb{E}[S_n] - S_n \geqslant \|b\|_{1,a} x) \leqslant - \frac{\|b\|_{1,a}^2}{\|b\|_{2,a}^2} \left(\frac{\|b\|_{2,a}^2 xt}{\|b\|_{1,a}} - \ell_c^* \left(\frac{\|b\|_{2,a}^2 t}{\|b\|_{1,a}} \right) \right),$$

$$\leqslant - \frac{\|b\|_{1,a}^2}{\|b\|_{2,a}^2} \left(\frac{\|b\|_{2,a}^2 (x-1) t}{\|b\|_{1,a}} + \log \left(1 + \frac{\|b\|_{2,a}^2 t}{\|b\|_{1,a}} \right) \right).$$

In view of the above inequality, it is necessary to assume that x belongs to $]0,1[$. By taking the optimal $t = \|b\|_{1,a} x / (\|b\|_{2,a}^2 (1-x))$ in this inequality, we find that

$$\log \mathbb{P}(\mathbb{E}[S_n] - S_n \geqslant \|b\|_{1,a} x) \leqslant - \frac{\|b\|_{1,a}^2}{\|b\|_{2,a}^2} \ell_c(x),$$

which clearly leads to (2.105). Finally, (2.106) immediately follows from (2.105), which completes the proof of Theorem 2.57. $\qquad\qquad\square$

2.8 McDiarmid's inequality

This section is devoted to the so-called McDiarmid inequality for functions of independent random variables. First of all, we recall the usual version of this inequality. Let $(E_1, d_1), \ldots, (E_n, d_n)$ be a finite sequence of separable metric spaces with respective finite diameters $c_1, \ldots, c_n$. Denote E^n the product space $E^n = E_1 \times \cdots \times E_n$.

Definition 2.60. A function f from E^n into $\mathbb{R}$ is said to be separately 1-Lipschitz if, for any x, y in E^n with $x = (x_1, \ldots, x_n)$ and $y = (y_1, \ldots, y_n)$,

$$|f(x_1, \ldots, x_n) - f(y_1, \ldots, y_n)| \leqslant \sum_{k=1}^{n} d_k(x_k, y_k). \tag{2.115}$$

Let $X_1, \ldots, X_n$ be a finite sequence of independent random variables such that the random vector $(X_1, \ldots, X_n)$ takes its values in E^n. McDiarmid's inequality says that for any separately 1-Lipschitz function f, the random variable

$$Z = f(X_1, \ldots, X_n) \tag{2.116}$$

satisfies the concentration inequality given, for any positive x, by

$$\mathbb{P}(Z - \mathbb{E}[Z] \geqslant x) \leqslant \exp\left(-\frac{2x^2}{C_n}\right) \quad \text{where} \quad C_n = \sum_{k=1}^{n} c_k^2. \tag{2.117}$$

This inequality was obtained by McDiarmid [17, 18]. We refer to Exercise 9 in Chapter 3 for the proof of this inequality, which uses a martingale method due to Yurinskii [23].

Remark 2.61. If the space E^n is countable and the distances $d_1, \ldots, d_n$ are defined, for all $1 \leqslant k \leqslant n$, by $d_k(x_k, y_k) = c_k$ if $x_k \neq y_k$, then (2.115) is equivalent, for any x, y in E^n with $x = (x_1, \ldots, x_n)$ and $y = (y_1, \ldots, y_n)$, to

$$|f(x_1, \ldots, x_n) - f(y_1, \ldots, y_n)| \leqslant \sum_{k=1}^{n} c_k \mathbb{I}_{x_k \neq y_k}.$$

It ensures that f is uniformly bounded. In that case, the optimal reals c_k in the above inequality are

$$c_k = \sup_{(x, y_k, z_k)} |f(x_1, \ldots, x_{k-1}, y_k, x_{k+1}, \ldots, x_n) - f(x_1, \ldots, x_{k-1}, z_k, x_{k+1}, \ldots, x_n)|.$$

We now propose an improvement of McDiarmid's inequality in the style of Delyon [10]: instead of assuming a uniform bound on each oscillation, we only assume a bound on the sum of squares. For all $1 \leqslant k \leqslant n$, denote by $\mathscr{F}^{(k)}$ the σ-algebra generated by $X_1, \ldots, X_n$ except X_k,

$$\mathscr{F}^{(k)} = \sigma(X_1, \ldots, X_{k-1}, X_{k+1}, \ldots, X_n).$$

Theorem 2.62. *Let $X_1, \ldots, X_n$ be a finite sequence of independent random variables and let Z be a measurable function of $X_1, \ldots, X_n$. Assume that for each $1 \leqslant k \leqslant n$, there exist two $\mathscr{F}^{(k)}$-measurable bounded random variables A_k and B_k such that*

$$A_k \leqslant Z \leqslant B_k \qquad a.s. \tag{2.118}$$

Then, for any positive x,

$$\mathbb{P}(Z \geqslant \mathbb{E}[Z] + x) \leqslant \exp\left(-\frac{2x^2}{D_n}\right) \quad \text{where} \quad D_n = \left\| \sum_{k=1}^{n} (B_k - A_k)^2 \right\|_\infty. \tag{2.119}$$

Remark 2.63. Assume that the space E^n is countable and $Z = f(X_1, \ldots, X_n)$ where f is a separately 1-Lipschitz function on E^n. For all $1 \leqslant k \leqslant n$, denote

$$A_k = \inf_{x_k \in E_k} f(X_1, \ldots, X_{k-1}, x_k, X_{k+1}, \ldots, X_n),$$

and

$$B_k = \sup_{x_k \in E_k} f(X_1, \ldots, X_{k-1}, x_k, X_{k+1}, \ldots, X_n).$$

Then, we clearly have

$$B_k - A_k \leqslant c_k \qquad \text{a.s.}$$

which shows that $D_n \leqslant C_n$. It means that Theorem 2.62 improves McDiarmid's inequality.

The proof is based on Log-Sobolev type inequalities, which have been widely developed by Ledoux [14]. The following lemma, due to Boucheron, Lugosi, and Massart [8], will be the main tool in the proof of our result.

Lemma 2.64. *Let $X_1, \ldots, X_n$ be a finite sequence of independent random variables and let Z be a measurable function of $X_1, \ldots, X_n$. Let $Z^{(1)}, \ldots, Z^{(n)}$ be any finite sequence of real bounded random variables such that, for each $1 \leqslant k \leqslant n$, $Z^{(k)}$ is $\mathscr{F}^{(k)}$-measurable. Denote by φ the function defined, for any real x, by $\varphi(x) = \exp(-x) + x - 1$. Then, for any positive t,*

$$\mathbb{E}[tZe^{tZ}] - \mathbb{E}[e^{tZ}]\log\big(\mathbb{E}[e^{tZ}]\big) \leqslant \sum_{k=1}^{n} \mathbb{E}\big[e^{tZ}\varphi(tZ - tZ^{(k)})\big]. \qquad (2.120)$$

Proof. We shall only prove Lemma 2.64 in the particular case $t = 1$, as the general case follows by multiplying Z by t. Let $\mathscr{F}_0$ be the trivial σ-algebra $\mathscr{F}_0 = \{\emptyset, \Omega\}$, and for all $1 \leqslant k \leqslant n$, denote $\mathscr{F}_k = \sigma(X_1, \ldots, X_k)$ and $Z_k = \log\big(\mathbb{E}[e^Z | \mathscr{F}_k]\big)$. We have by a standard telescopic argument

$$\mathbb{E}[Ze^Z] - \mathbb{E}[e^Z]\log\mathbb{E}[e^Z] = \sum_{k=1}^{n} \mathbb{E}\big[e^Z(Z_k - Z_{k-1})\big]. \qquad (2.121)$$

The right-hand side of (2.121) can be rewritten as

$$\mathbb{E}\big[e^Z(Z_k - Z_{k-1})\big] = \mathbb{E}\big[e^{Z^{(k)}} e^{Z - Z^{(k)}}(Z_k - Z_{k-1})\big].$$

Then, it follows from the Young type inequality $xy \leqslant x\log(x) - x + \exp(y)$ with $x > 0$ and y in $\mathbb{R}$, applied to $x = \exp(Z - Z^{(k)})$ and $y = Z_k - Z_{k-1}$, that

$$\mathbb{E}\big[e^Z(Z_k - Z_{k-1})\big] \leqslant \mathbb{E}\big[e^Z(Z - Z^{(k)}) - e^Z + e^{Z^{(k)} + Z_k - Z_{k-1}}\big].$$

However, the independence of the underlying random variables implies that the random variable $\mathbb{E}[e^{Z^{(k)}} | \mathscr{F}_k]$ is $\mathscr{F}_{k-1}$-measurable. Hence,

$$\mathbb{E}\left[e^{Z^{(k)}+Z_k-Z_{k-1}}\right] = \mathbb{E}\left[\mathbb{E}[e^{Z^{(k)}}|\mathscr{F}_k]e^{Z_k-Z_{k-1}}\right],$$
$$= \mathbb{E}\left[\mathbb{E}[e^{Z^{(k)}}|\mathscr{F}_{k-1}]e^{Z_k-Z_{k-1}}\right],$$
$$= \mathbb{E}\left[e^{Z^{(k)}}\right].$$

Consequently,

$$\mathbb{E}[e^{Z}(Z_k - Z_{k-1})] \leqslant \mathbb{E}\left[e^{Z}(Z - Z^{(k)}) - e^{Z} + e^{Z^{(k)}}\right],$$

which, together with (2.121), leads to (2.120). $\qquad\qquad\qquad\square$

Proof of Theorem 2.62. For any positive t, denote $F(t) = \mathbb{E}[\exp(tZ)]$ and $L(t) = \log(F(t))$. It follows from Lemma 2.64 that, for any positive t,

$$F(t)(tL'(t) - L(t)) \leqslant \sum_{k=1}^{n} \mathbb{E}[\exp(tZ)\varphi(tZ - tZ^{(k)})]. \qquad (2.122)$$

Now, the function φ is convex. Since the random variable Z belongs to $[A_k, B_k]$ almost surely, it implies that

$$\varphi(tZ - tZ^{(k)}) \leqslant \max\left(\varphi(tA_k - tZ^{(k)}), \varphi(tB_k - tZ^{(k)})\right).$$

Consequently, it is natural to choose $Z^{(k)}$ in such a way that

$$\varphi(tA_k - tZ^{(k)}) = \varphi(tB_k - tZ^{(k)}).$$

The solution of this equation is given by

$$Z^{(k)} = \frac{1}{t}\left(tA_k + \log(tC_k) - \log(1 - \exp(-tC_k))\right)$$

where $C_k = B_k - A_k$. For this choice of $Z^{(k)}$, we find that for any positive t,

$$\varphi(tZ - tZ^{k}) \leqslant \ell(tC_k) \qquad (2.123)$$

where the function ℓ is defined by $\ell(0) = 0$ and, for any $x \neq 0$,

$$\begin{aligned}
\ell(x) &= \frac{x}{1 - \exp(-x)} + \log\left(\frac{1 - \exp(-x)}{x}\right) - 1, \\
&= \frac{x\exp(x)}{\exp(x) - 1} - x + \log\left(\frac{\exp(x) - 1}{x}\right) - 1, \\
&= \frac{x}{\exp(x) - 1} + \log\left(\frac{\exp(x) - 1}{x}\right) - 1. \qquad (2.124)
\end{aligned}$$

One can observe that $\ell(-x) = \ell(x)$. We now claim that, for any real x,

$$\ell(x) \leqslant \frac{x^2}{8}. \tag{2.125}$$

As a matter of fact, denote by h_r the function defined, for any r in $[0,1]$ and for any real x, by

$$h_r(x) = \log\left(re^{(1-r)x} + (1-r)e^{-rx}\right).$$

The function h_r is convex with respect to x and concave with respect to r. Now, for any $x \neq 0$, its maximum with respect to r is attained for

$$r_x = \frac{e^x - x - 1}{xe^x - x}.$$

Therefore,

$$\sup_{r\in[0,1]} h_r(x) = h_{r_x}(x) = \ell(x). \tag{2.126}$$

However, for any r in $[0,1]$, h_r is the log-Laplace transform of a centered random variable ε with two-value distribution given by $\mathbb{P}(\varepsilon = 1 - r) = r$ and $\mathbb{P}(\varepsilon = -r) = 1 - r$. We immediately deduce from Lemma 2.19 that, for any r in $[0,1]$ and for any real x, $h_r(x) \leqslant x^2/8$. Consequently, (2.126) clearly leads to (2.125). Hereafter, it follows from the conjunction of (2.122), (2.123), and (2.125) that, for any positive t,

$$F(t)(tL'(t) - L(t)) \leqslant \frac{t^2}{8}\, \mathbb{E}\left[\exp(tZ) \sum_{k=1}^{n} C_k^2\right] \leqslant \frac{t^2}{8} D_n F(t),$$

which implies that

$$\frac{tL'(t) - L(t)}{t^2} \leqslant \frac{D_n}{8}.$$

Integrating this inequality, we obtain that, for any positive t,

$$\frac{L(t)}{t} - L'(0) \leqslant \frac{tD_n}{8},$$

leading to

$$\log \mathbb{E}\left[\exp(tZ)\right] \leqslant t\mathbb{E}[Z] + D_n\frac{t^2}{8}. \tag{2.127}$$

Finally, we infer from Markov's inequality and (2.127) that for any positive x and for any positive t,

$$\log \mathbb{P}(Z \geqslant \mathbb{E}[Z] + x) \leqslant -tx + \frac{D_n t^2}{8}.$$

By taking the optimal value $t = 4x/D_n$ in this inequality, we immediately obtain (2.119), which achieves the proof of Theorem 2.62. $\qquad\square$

2.9 Complements and Exercises

Exercise 1. Let X be a real-valued random variable with finite Laplace transform on a right neighborhood of the origin. Denote $L_X(t) = \log \mathbb{E}[\exp(tX)]$. Let Ψ_X be the function defined, for any $x \geqslant 0$, by

$$\Psi_X(x) = \inf_{t>0}\left(t^{-1}(L_X(t) + x)\right).$$

Let L_X^* be the Legendre-Fenchel dual of L_X. Prove that $(\Psi_X(x) < y)$ if and only if $(L_X^*(y) > x)$. Deduce that $\Psi_X = L_X^{*-1}$. Moreover, let X and Y be real-valued random variables with finite Laplace transforms on a right neighborhood of the origin. Prove that

$$L_{X+Y}^{*-1} \leqslant L_X^{*-1} + L_Y^{*-1}.$$

Hint: Apply the Hölder inequality to the product $\exp(tX)\exp(tY)$.

Exercise 2 (A reversed Bernstein's type inequality). Let $X_1, \ldots, X_n$ be a finite sequence of independent random variables, satisfying (2.2) with $c = 1$ and $\mathbb{E}[S_n] = 0$. Let v_n be defined by (2.1). Prove that for any positive x,

$$\mathbb{P}\left(S_n > n\left(\frac{1}{2} + \frac{\log(1+x)}{2x}\right)\left(\sqrt{2v_n x + x^2} + x\right)\right) \leqslant \exp(-nx).$$

Hint: Use (2.16) and choose t in such a way that $v_n t^2 = 2x(1-t)$. Compare this inequality with the inequalities of Theorem 2.1.

Exercise 3 (A Bernstein's type inequality for symmetric random variables). Let $X_1, \ldots, X_n$ be a finite sequence of independent random variables with symmetric distribution, satisfying the two-sided Bernstein's condition (2.18) with $c = 1$.

1) Prove that, for any t in $]0, 1[$,

$$\log \mathbb{E}\left[\exp(tS_n)\right] \leqslant n\log\left(1 + \frac{v_n t^2}{2(1-t^2)}\right).$$

2) Choosing t into (2.7), in such a way that $v_n t = x(1-t^2)$, prove that for any positive x,

$$\mathbb{P}(S_n \geqslant nx) \leqslant \left(1 + \gamma_n(x)\right)^n \exp\left(-2n\gamma_n(x)\right)$$

where

$$\gamma_n(x) = \frac{x^2}{v_n + \sqrt{v_n^2 + 4x^2}}.$$

Exercise 4 (Direct proof of Bennett's inequality). Let $X_1, \ldots, X_n$ be a finite sequence of independent random variables with values in $]-\infty, 1]$ and finite variances. Assume that $\mathbb{E}[S_n] = 0$.

1) Let X be a centered random variable with values in $]-\infty, 1]$ and finite variance v. Prove that, for any positive t, $\mathbb{E}[\exp(tX)] \leqslant 1 + v(\exp(t) - t - 1)$.

2) Use Lemma 2.6 to prove that, for any positive t,

$$\log \mathbb{E}\big[\exp(tS_n)\big] \leqslant n\log\big(1 + v_n(\exp(t) - t - 1)\big) \leqslant nv_n\big(\exp(t) - t - 1\big).$$

3) Prove that, for any positive x,

$$\mathbb{P}(S_n \geqslant nx) \leqslant \big(1 + x - v_n\log(1 + x/v_n)\big)^n \exp\big(-nx\log(1 + x/v_n)\big).$$

4) Deduce from the above result that, for any positive x,

$$\mathbb{P}(S_n \geqslant nx) \leqslant \exp\big(-nv_n h(x/v_n)\big)$$

where $h(x) = (1 + x)\log(1 + x) - x$.

Exercise 5 (Massart's inequality, [15]). Let S_n be a random variable with Binomial $\mathscr{B}(n,p)$ distribution. Use Theorem 2.28 to prove that, for any x in $[0, 1 - p]$,

$$\mathbb{P}(S_n - np \geqslant nx) \leqslant \exp\left(-\frac{nx^2}{2(p + x/3)(1 - p - x/3)}\right).$$

Exercise 6. Let $X_1, \ldots, X_n$ be a finite sequence of independent random variables such that $X_k \leqslant b$ a.s. for some positive constant b. Let S_n and v_n be defined as in (2.1). Prove that, for any positive t,

$$\mathbb{P}\big(S_n \geqslant \sqrt{2nv_n t} + \max(0, b - v_n/b)t/3\big) \leqslant \exp(-t).$$

Deduce that

$$\mathbb{P}\big(S_n \geqslant \sqrt{2nv_n t} + bt/3\big) \leqslant \exp(-t).$$

Hint: Use the first part of Theorem 2.28.

Exercise 7 (Hoeffding Binomial inequality for nonnegative random variables). Let $X_1, \ldots, X_n$ be nonnegative independent random variables. Denote

$$E_n = \frac{1}{n}\sum_{k=1}^{n}\mathbb{E}[X_k] \qquad \text{and} \qquad D_n = \frac{1}{n}\sum_{k=1}^{n}\mathbb{E}[X_k^2].$$

Prove that, for any x in $]0, 1[$,

$$\mathbb{P}\big(S_n \leqslant x\mathbb{E}[S_n]\big) \leqslant \exp\left(-\frac{n}{D_n}\left((D_n - E_n^2 x)\log\left(\frac{D_n - E_n^2 x}{D_n - E_n^2}\right) + E_n^2 x\log x\right)\right).$$

Exercise 8. Let $\varepsilon_1, \ldots, \varepsilon_n$ be a finite sequence of independent random variables sharing the same Exponential $\mathscr{E}(1)$ distribution. Let $a_1, \ldots, a_n$ be a finite sequence of real numbers. For all $1 \leqslant k \leqslant n$, let $X_k = a_k(\varepsilon_k - 1)$. Prove that, for any positive x,

$$\mathbb{P}\big(S_n \geqslant \|a\|_2\sqrt{2x} + \max(0, a_1, a_2, \ldots, a_n)x\big) \leqslant \exp(-x).$$

Exercise 9 (Krafft's inequality, [13]). Let $X_1, \ldots, X_n$ be a finite sequence of independent random variables with values in $[0,1]$. Prove that, for any positive x,

$$\mathbb{P}(S_n \geqslant \mathbb{E}[S_n] + nx) \leqslant \exp\left(-2nx^2 - \frac{4}{9}nx^4\right).$$

Exercise 10 (Weighted sums). Let $X_1, \ldots, X_n$ be a finite sequence of independent random variables.

1) Assume that, for all $1 \leqslant k \leqslant n$, $X_k = b_k Z_k$ where b_k is a real number, not necessarily positive, and $Z_1, \ldots, Z_n$ is a finite sequence of independent random variables such that, for all $1 \leqslant k \leqslant n$, Z_k has the $\Gamma(a_k, 1)$ distribution where $a_k > 0$. Assume that $\beta = \max(b_1, \ldots, b_n) > 0$. Prove that, for any positive x,

$$\mathbb{P}(S_n - \mathbb{E}[S_n] \geqslant \|b\|_{2,a}^2 x) \leqslant \exp\left(-\frac{\|b\|_{2,a}^2}{\beta^2}\Big(\beta x - \log(1 + \beta x)\Big)\right).$$

where $\|b\|_{2,a}$ is defined by (2.102).

2) Assume that, for all $1 \leqslant k \leqslant n$, $X_k = c_k \varepsilon_k$ where c_k is a positive real number and $\varepsilon_1, \ldots, \varepsilon_n$ is a finite sequence of independent random variables sharing the same Bernoulli $\mathscr{B}(p)$ distribution. If $0 < p \leqslant 1/2$, prove that for any x in $[0, p]$,

$$\mathbb{P}(S_n \leqslant \|c\|_1 x) \leqslant \exp\left(-\frac{\|c\|_1^2}{\|c\|_2^2}\left(x\log\left(\frac{x}{p}\right) + (1 - x)\log\left(\frac{1 - x}{1 - p}\right)\right)\right).$$

References

1. Antonov, S. N.: Probability inequalities for series of independent random variables. Teor. Veroyatnost. i Primenen. **24**, 632–636 (1979)
2. Bennett, G.: Probability inequalities for the sum of independent random variables. J. Amer. Statist. Assoc. **57**, 33–45 (1962)
3. Bennett, G.: On the probability of large deviations from the expectation for sums of bounded, independent random variables. Biometrika **50**, 528–635 (1963)
4. Bentkus, V.: On Hoeffding's inequalities. Ann. Probab. **32**, 1650–1673 (2004)
5. Bentkus, V.: An inequality for tail probabilities of martingales with differences bounded from one side. J. Theoret. Probab. **16**, 161–173 (2003)
6. Bernstein, S. N.: Theory of Probability, Moscow (1927)
7. Boucheron, S., Lugosi, G. and Massart, P.: Concentration inequalities. Oxford University Press, Oxford (2013)
8. Boucheron, S., Lugosi, G. and Massart, P.: A sharp concentration inequality with applications. Random Structures Algorithms. **16**, 277–29 (2000)
9. Del Moral, P. and Rio, E.: Concentration inequalities for mean field particle models. Ann. Appl. Probab. **21**, 1017–1052 (2011)
10. Delyon, B.: Concentration inequalities for the spectral measure of random matrices. Electron. Commun. Probab. **15**, 549–561 (2010)
11. Hoeffding, W.: Probability inequalities for sums of bounded random variables. J. Amer. Statist. Assoc. **58**, 13–30 (1963)

12. Kearns, M. J. and Saul, L. K.: Large deviation methods for approximate probabilistic inference. Proceedings of the 14th Conference on Uncertaintly in Artifical Intelligence, San-Francisco, 311–319 (1998)
13. Krafft, O.: A note on exponential bounds for binomial probabilities. Ann. Inst. Stat. Math. **21**, 219–220 (1969)
14. Ledoux, M.: Isoperimetry and Gaussian analysis. Lectures on probability theory and statistic. Lecture Notes in Math. **1648**, 165–294 (1996)
15. Massart, P.: The tight constant in the Dvoretzky-Kiefer-Wolfowitz inequality. Ann. Probab. **18**, 1269–1283 (1990)
16. Maurer, A.: A bound on the deviation probability for sums of non-negative random variables. J. Inequal. Pure Appl. Math. **4**, Article 15 (2003)
17. McDiarmid, C.: On the method of bounded differences. Surveys in combinatorics, London Mathematical Society lecture notes series **141** 148–188 (1989)
18. McDiarmid, C.: Concentration. Probabilistic methods for algorithmic discrete mathematics. Springer-Verlag, Berlin **16** 195–248 (1998)
19. Pinelis, I.: On the Bennett-Hoeffding inequality. Ann. Inst. Henri Poincaré Probab. Stat. **50**, 15–27 (2014)
20. Rio, E.: On McDiarmid's concentration inequality. Electron. Commun. Probab. **18**, 1–11 (2013)
21. Rio, E.: Extensions of the Hoeffding-Azuma inequalities. Electron. Commun. Probab. **18**, 1–6 (2013)
22. Rio, E.: Inégalités exponentielles et inégalités de concentration. Hal, cel-00702524 (2012)
23. Yurinskii, V. V.: Exponential bounds for large deviations, Teor. Veroyatnost. i Primenen. **19**, 152–154 (1974)

Chapter 3
Concentration inequalities for martingales

3.1 Azuma-Hoeffding inequalities

Throughout this section, (M_n) is a square integrable martingale with bounded differences, adapted to a filtration $\mathbb{F} = (\mathscr{F}_n)$, such that $M_0 = 0$. Its increasing process is defined by $\langle M \rangle_0 = 0$ and, for all $n \geqslant 1$,

$$\langle M \rangle_n = \sum_{k=1}^{n} \mathbb{E}[(M_k - M_{k-1})^2 | \mathscr{F}_{k-1}]. \tag{3.1}$$

In all the sequel, we shall denote $\Delta M_n = M_n - M_{n-1}$ and

$$V_n = \langle M \rangle_n - \langle M \rangle_{n-1} = \mathbb{E}[(M_n - M_{n-1})^2 | \mathscr{F}_{n-1}]. \tag{3.2}$$

Hoeffding [18] realized that Theorem 2.16 holds also true for martingales (*see* [18], p. 18). More precisely, assume that (M_n) is a martingale such that, for all $1 \leqslant k \leqslant n$, one can find two constants $a_k < b_k$ satisfying $a_k \leqslant \Delta M_k \leqslant b_k$ almost surely. Then, for any positive x,

$$\mathbb{P}(M_n \geqslant x) \leqslant \exp\left(-\frac{2x^2}{D_n}\right) \quad \text{where} \quad D_n = \sum_{k=1}^{n} (b_k - a_k)^2. \tag{3.3}$$

Later, Azuma [1] gave a complete proof of (3.3) in the symmetric case $a_k = -b_k$. Our goal in this section is to provide new versions of Azuma-Hoeffding type inequalities. However, instead of considering deterministic bounds on the increments, we will only assume that, for all $1 \leqslant k \leqslant n$, one can find a negative bounded random variable A_k and a positive bounded random variable B_k such that the couple (A_k, B_k) is $\mathscr{F}_{k-1}$-measurable and it satisfies

$$A_k \leqslant \Delta M_k \leqslant B_k \qquad \text{a.s.} \tag{3.4}$$

© The Authors 2015

B. Bercu et al., *Concentration Inequalities for Sums and Martingales*,
SpringerBriefs in Mathematics, DOI 10.1007/978-3-319-22099-4_3

Our strategy is motivated by the widespread example of martingale transforms given below. Let (ε_n) be a sequence of random variables, adapted to $\mathbb{F}$, such that for all $k \geqslant 1$, $\mathbb{E}[\varepsilon_k | \mathscr{F}_{k-1}] = 0$ and $a_k \leqslant \varepsilon_k \leqslant b_k$ almost surely, for two constants $a_k < b_k$. In addition, let (ϕ_n) be a sequence of positive and bounded random variables, adapted to $\mathbb{F}$. For all $n \geqslant 1$, denote

$$M_n = \sum_{k=1}^{n} \phi_{k-1} \varepsilon_k. \tag{3.5}$$

The sequence (M_n) is commonly called a martingale transform. One can obviously see that (3.4) holds true with $A_k = a_k \phi_{k-1}$ and $B_k = b_k \phi_{k-1}$. In Subsection 3.1.1, we extend the so-called Kearns-Saul's inequality [19] to martingales with differences bounded from above. Subsection 3.1.2 is devoted to martingales satisfying symmetric boundedness assumptions, while Subsection 3.1.3 deals with several improvements of Azuma-Hoeffding's inequality, including Van de Geer's inequality [21].

3.1.1 Martingales with differences bounded from above

Throughout this subsection, we assume that (M_n) is a martingale satisfying, for all $1 \leqslant k \leqslant n$, the one-sided boundedness condition

$$\Delta M_k \leqslant B_k \qquad \text{a.s.} \tag{3.6}$$

where B_k is a positive and bounded $\mathscr{F}_{k-1}$-measurable random variable. Under this assumption, Theorem 2.33 can be extended to martingales as follows.

Theorem 3.1. *Let (M_n) be a square integrable martingale satisfying (3.6) and let (V_n) be the sequence given by (3.2). Denote by φ the function*

$$\varphi(v) = \begin{cases} \dfrac{1-v^2}{|\log(v)|} & \text{if } v < 1, \\ 2v & \text{if } v \geqslant 1. \end{cases} \tag{3.7}$$

Then, for any positive x and y,

$$\mathbb{P}(M_n \geqslant x, \mathscr{A}_n \leqslant y) \leqslant \exp\left(-\frac{x^2}{y}\right) \qquad \text{where} \qquad \mathscr{A}_n = \sum_{k=1}^{n} B_k^2 \varphi\left(\frac{V_k}{B_k^2}\right). \tag{3.8}$$

Consequently,

$$\mathbb{P}(M_n \geqslant x, 6\langle M \rangle_n + \mathscr{C}_n \leqslant y) \leqslant \exp\left(-\frac{3x^2}{y}\right) \tag{3.9}$$

where

$$\mathscr{C}_n = \sum_{k=1}^{n} \left(B_k - \frac{V_k}{B_k} \right)_+^2.$$

In addition, for any positive x,

$$\mathbb{P}(M_n \geqslant x) \leqslant \exp\left(-\frac{x^2}{\|\mathscr{A}_n\|_\infty} \right) \leqslant \exp\left(-\frac{3x^2}{\|6\langle M\rangle_n + \mathscr{C}_n\|_\infty} \right). \tag{3.10}$$

Remark 3.2. A concentration inequality similar to (3.9) can be found in the recent contribution of Fan, Grama, and Liu [16].

The proof of Theorem 3.1 relies on the following lemma which makes use of a suitable exponential supermartingale.

Lemma 3.3. *Let (M_n) be a martingale such that $M_0 = 0$ and let p be any real with $p > 1$. Assume that, for all $n \geqslant 1$, one can find a nonnegative and $\mathscr{F}_{n-1}$-measurable random variable W_n such that, for any positive t,*

$$\log\big(\mathbb{E}[\exp(t\Delta M_n) \mid \mathscr{F}_{n-1}] \big) \leqslant \frac{t^p}{p} W_n \qquad a.s. \tag{3.11}$$

Then, for any positive x and y,

$$\mathbb{P}(M_n \geqslant x, \mathscr{W}_n \leqslant y) \leqslant \exp\left(-\frac{1}{q} \frac{x^q}{y^{q-1}} \right) \tag{3.12}$$

where

$$\mathscr{W}_n = \sum_{k=1}^{n} W_k$$

and $q = p/(p-1)$ is the Hölder conjugate exponent of p.

Proof. For any real t and for all $n \geqslant 0$, denote

$$V_n(t) = \exp\left(tM_n - \frac{t^p}{p} \mathscr{W}_n \right)$$

with $V_0(t) = 1$. We claim that for any real t, $(V_n(t))$ is a positive supermartingale such that $\mathbb{E}[V_n(t)] \leqslant 1$. As a matter of fact, for any real t and for all $n \geqslant 0$, we clearly have

$$V_n(t) = V_{n-1}(t) \exp\left(t\Delta M_n - \frac{t^p}{p} W_n \right).$$

It follows from condition (3.11) that, for any real t,

$$\mathbb{E}[V_n(t)|\mathscr{F}_{n-1}] = V_{n-1}(t)\mathbb{E}\big[\exp(t\Delta M_n)|\mathscr{F}_{n-1}\big]\exp\Big(-\frac{t^p}{p}W_n\Big),$$

$$\leqslant V_{n-1}(t)\exp\Big(\frac{t^p}{p}W_n\Big)\exp\Big(-\frac{t^p}{p}W_n\Big) = V_{n-1}(t).$$

Consequently, $(V_n(t))$ is a positive supermartingale. By taking the expectation on both sides of the above inequality, we obtain that for all $n \geqslant 1$, $\mathbb{E}[V_n(t)] \leqslant \mathbb{E}[V_{n-1}(t)]$, which leads to $\mathbb{E}[V_n(t)] \leqslant 1$. We are now in position to prove Lemma 3.3. For any positive x and y, let

$$A_n = \Big\{M_n \geqslant x, \mathscr{W}_n \leqslant y\Big\}.$$

By Markov's inequality, we have for any positive t,

$$\mathbb{P}(A_n) \leqslant \mathbb{E}\Big[\exp\Big(tM_n - tx\Big)\mathrm{I}_{A_n}\Big],$$

$$\leqslant \mathbb{E}\Big[\exp\Big(tM_n - \frac{t^p}{p}\mathscr{W}_n\Big)\exp\Big(\frac{t^p}{p}\mathscr{W}_n - tx\Big)\mathrm{I}_{A_n}\Big],$$

$$\leqslant \exp\Big(\frac{t^p y}{p} - tx\Big)\mathbb{E}[V_n(t)],$$

$$\leqslant \exp\Big(\frac{t^p y}{p} - tx\Big).$$

Hence, by taking the optimal value $t = (x/y)^{q/p}$ in the above inequality, we find that

$$\mathbb{P}(A_n) \leqslant \exp\Big(\frac{y}{p}\Big(\frac{x}{y}\Big)^q - x\Big(\frac{x}{y}\Big)^{q/p}\Big) = \exp\Big(-\frac{1}{q}y\Big(\frac{x}{y}\Big)^q\Big),$$

which is exactly what we wanted to prove. $\qquad\qquad\square$

Proof of Theorem 3.1. For all $n \geqslant 1$, denote

$$Z_n = \frac{\Delta M_n}{B_n}.$$

It clearly follows from condition (3.6) that, for all $n \geqslant 1$, $Z_n \leqslant 1$ a.s. Moreover, we have, almost surely

$$\mathbb{E}[Z_n|\mathscr{F}_{n-1}] = 0 \qquad \text{and} \qquad \mathbb{E}[Z_n^2|\mathscr{F}_{n-1}] = \frac{V_n}{B_n^2}.$$

Hence, according to Lemma 2.36, we have for all $n \geqslant 1$ and for any positive t,

$$\log\mathbb{E}[\exp(tZ_n)|\mathscr{F}_{n-1}] \leqslant \frac{1}{4}\varphi\Big(\frac{V_n}{B_n}\Big)t^2 \qquad \text{a.s.}$$

which ensures that

$$\log \mathbb{E}[\exp(t\Delta M_n)|\mathscr{F}_{n-1}] \leqslant \frac{1}{4}B_n^2\varphi\left(\frac{V_n}{B_n}\right)t^2 \qquad \text{a.s.}$$

Consequently, we deduce from Lemma 3.3 with $p = 2$ that

$$\mathbb{P}(M_n \geqslant x, \mathscr{W}_n \leqslant y) \leqslant \exp\left(-\frac{x^2}{2y}\right) \tag{3.13}$$

where

$$\mathscr{W}_n = \sum_{k=1}^{n}\frac{1}{2}B_k^2\varphi\left(\frac{V_k}{B_k}\right) = \frac{1}{2}\mathscr{A}_n,$$

which clearly implies (3.8). We obtain (3.10) by taking the value $y = \|\mathscr{A}_n\|_\infty$ into (3.8). Furthermore, we infer from Lemma 2.37 that

$$\mathscr{A}_n = \sum_{k=1}^{n} B_k^2\varphi\left(\frac{V_k}{B_k^2}\right) \leqslant \frac{1}{3}\left(6\mathscr{V}_n + \mathscr{C}_n\right) \tag{3.14}$$

where

$$\mathscr{V}_n = \sum_{k=1}^{n} V_k = \sum_{k=1}^{n} \langle M\rangle_k - \langle M\rangle_{k-1} = \langle M\rangle_n$$

and

$$\mathscr{C}_n = \sum_{k=1}^{n}\left(B_k - \frac{V_k}{B_k}\right)_+^2.$$

Finally, (3.8) and (3.14) imply (3.9), which completes the proof of Theorem 3.1. $\square$

3.1.2 Symmetric conditions for bounded difference martingales

This subsection deals with the situation where the martingale (M_n) satisfies, for all $1 \leqslant k \leqslant n$, the symmetric boundedness condition

$$|\Delta M_k| \leqslant B_k \qquad \text{a.s.} \tag{3.15}$$

where B_k is a positive and bounded $\mathscr{F}_{k-1}$-measurable random variable. It is inspired by the original work of Azuma [1].

Theorem 3.4. *Let (M_n) be a square integrable martingale such that $M_0 = 0$. Assume that (M_n) satisfies (3.15). Then, for any positive x and y,*

$$\mathbb{P}(M_n \geqslant x, 5\langle M\rangle_n + \mathscr{B}_n \leqslant y) \leqslant \exp\left(-\frac{3x^2}{y}\right) \tag{3.16}$$

where

$$\mathscr{B}_n = \sum_{k=1}^{n} B_k^2.$$

Consequently, for any positive x,

$$\mathbb{P}(M_n \geqslant x) \leqslant \exp\left(-\frac{3x^2}{\|5\langle M\rangle_n + \mathscr{B}_n\|_\infty}\right). \tag{3.17}$$

Remark 3.5. According to condition (3.15),

$$\langle M\rangle_n = \sum_{k=1}^{n} \mathbb{E}[(M_k - M_{k-1})^2|\mathscr{F}_{k-1}] \leqslant \sum_{k=1}^{n} B_k^2 = \mathscr{B}_n.$$

Hence, (3.17) implies that, for any positive x,

$$\mathbb{P}(M_n \geqslant x) \leqslant \exp\left(-\frac{x^2}{2\|\mathscr{B}_n\|_\infty}\right), \tag{3.18}$$

which improves Azuma's inequality.

Proof of Theorem 3.4. It follows from condition (3.15) that, for all $1 \leqslant k \leqslant n$, $V_k \leqslant B_k^2$. Hence,

$$\begin{aligned}
\mathscr{C}_n &= \sum_{k=1}^{n} \left(B_k - \frac{V_k}{B_k}\right)_+^2 \leqslant \sum_{k=1}^{n} B_k^2 + \sum_{k=1}^{n} V_k\left(-2 + \frac{V_k}{B_k^2}\right), \\
&\leqslant \sum_{k=1}^{n} B_k^2 - \sum_{k=1}^{n} V_k \leqslant \sum_{k=1}^{n} B_k^2 = \mathscr{B}_n.
\end{aligned}$$

Hence, inequality (3.9) immediately leads to (3.16). Finally, we obtain (3.17) by taking the value $y = \|5\langle M\rangle_n + \mathscr{B}_n\|_\infty$ into (3.16). $\qquad\square$

3.1.3 Asymmetric conditions for bounded difference martingales

We now focus our attention on asymmetric boundedness conditions. As in Van de Geer [21], we assume that (M_n) satisfies, for all $1 \leqslant k \leqslant n$, the asymmetric boundedness condition

$$A_k \leqslant \Delta M_k \leqslant B_k \qquad \text{a.s.} \tag{3.19}$$

where the couple (A_k, B_k) is $\mathscr{F}_{k-1}$-measurable and A_k is a negative and bounded random variable, while B_k is a positive and bounded random variable.

Theorem 3.6. *Let (M_n) be a square integrable martingale such that $M_0 = 0$. Assume that (M_n) satisfies (3.19). Then, for any positive x and y,*

$$\mathbb{P}(M_n \geqslant x, 2\langle M \rangle_n + \mathscr{D}_n \leqslant y) \leqslant \exp\left(-\frac{3x^2}{y}\right) \tag{3.20}$$

where

$$\mathscr{D}_n = \sum_{k=1}^{n}(B_k - A_k)^2.$$

Consequently, for any positive x,

$$\mathbb{P}(M_n \geqslant x) \leqslant \exp\left(-\frac{3x^2}{\|2\langle M \rangle_n + \mathscr{D}_n\|_\infty}\right). \tag{3.21}$$

Remark 3.7. The convexity of the square function implies that, for all $1 \leqslant k \leqslant n$,

$$\Delta M_k^2 \leqslant (A_k + B_k)\Delta M_k - A_k B_k \qquad \text{a.s.}$$

By taking the conditional expectation on both sides of the above inequality, we obtain that, for all $1 \leqslant k \leqslant n$,

$$V_k = \mathbb{E}[\Delta M_k^2 | \mathscr{F}_{k-1}] \leqslant -A_k B_k \leqslant \frac{1}{4}(B_k - A_k)^2 \qquad \text{a.s.} \tag{3.22}$$

which ensures that

$$\langle M \rangle_n = \sum_{k=1}^{n} \mathbb{E}[(M_k - M_{k-1})^2 | \mathscr{F}_{k-1}] \leqslant \frac{1}{4}\sum_{k=1}^{n}(B_k - A_k)^2 = \frac{1}{4}\mathscr{D}_n. \tag{3.23}$$

Consequently (3.20) implies that, for any positive x and y,

$$\mathbb{P}(M_n \geqslant x, \mathscr{D}_n \leqslant y) \leqslant \exp\left(-\frac{2x^2}{y}\right). \tag{3.24}$$

This inequality, which is due to Van de Geer [21], leads to

$$\mathbb{P}(M_n \geqslant x) \leqslant \exp\left(-\frac{2x^2}{\|\mathscr{D}_n\|_\infty}\right). \tag{3.25}$$

By contrast, the classical Azuma-Hoeffding inequality for martingales states that, for any positive x,

$$\mathbb{P}(M_n \geqslant x) \leqslant \exp\left(-\frac{2x^2}{D_n}\right) \qquad \text{where} \qquad D_n = \sum_{k=1}^{n}(b_k - a_k)^2 \tag{3.26}$$

where, for all $1 \leqslant k \leqslant n$, $a_k < b_k$ are two constants such that $a_k \leqslant \Delta M_k \leqslant b_k$ a.s. One can realize that Azuma-Hoeffding's inequality is less efficient than (3.25).

Proof of Theorem 3.6. We already saw in inequality (3.22) that, for all $1 \leqslant k \leqslant n$, $V_k \leqslant -A_k B_k$ a.s. Hence,

$$6\langle M\rangle_n + \mathscr{C}_n \leqslant \sum_{k=1}^{n}\left(B_k^2 + 4V_k + \frac{V_k^2}{B_k^2}\right) \leqslant \sum_{k=1}^{n}\left(2V_k + B_k^2 - 2A_kB_k + A_k^2\right),$$

which means that

$$6\langle M\rangle_n + \mathscr{C}_n \leqslant 2\langle M\rangle_n + \mathscr{D}_n.$$

This upper bound, together with (3.9), immediately leads to Theorem 3.6. $\qquad\square$

Finally, we complete this subsection by an extension of Kearns-Saul inequality for martingales.

Theorem 3.8. *Let (M_n) be a square integrable martingale such that $M_0 = 0$. Assume that (M_n) satisfies (3.19). Then, for any positive x and y,*

$$\mathbb{P}(M_n \geqslant x, \mathscr{P}_n \leqslant y) \leqslant \exp\left(-\frac{x^2}{y}\right) \tag{3.27}$$

where

$$\mathscr{P}_n = \sum_{k=1}^{n} B_k^2 \varphi\left(-\frac{A_k}{B_k}\right).$$

Consequently, for any positive x,

$$\mathbb{P}(M_n \geqslant x) \leqslant \exp\left(-\frac{x^2}{\|\mathscr{P}_n\|_\infty}\right). \tag{3.28}$$

Proof. We deduce from inequality (3.22) and the fact that the function φ is increasing that

$$\mathscr{A}_n = \sum_{k=1}^{n} B_k^2 \varphi\left(\frac{V_k}{B_k^2}\right) \leqslant \sum_{k=1}^{n} B_k^2 \varphi\left(-\frac{A_k}{B_k}\right) = \mathscr{P}_n.$$

Now, Theorem 3.8 clearly follows from (3.8) via this upper bound.

Remark 3.9. One can observe that Theorem 3.8 outperforms Theorem 3.6.

3.2 Freedman and Fan-Grama-Liu inequalities

This section is devoted to martingales with differences uniformly bounded from above. Hence, we assume that (M_n) is a martingale satisfying, for all $1 \leqslant k \leqslant n$, the one-sided boundedness condition

$$\Delta M_k \leqslant b \qquad \text{a.s.} \tag{3.29}$$

where b is a positive real number. Our goal is to extend the results of Sections 2.3 and 2.4 to martingales. We start by an extension of Theorem 2.24 to martingales, due to Fan, Grama, and Liu [15].

Theorem 3.10. *Let (M_n) be a square integrable martingale such that $M_0 = 0$. Assume that (M_n) satisfies (3.29). Then, for any x in $[0,b]$ and for any positive y,*

$$\mathbb{P}(M_n \geqslant nx, \langle M \rangle_n \leqslant ny)$$

$$\leqslant \exp\left(-n\left(\frac{y+bx}{y+b^2}\log\left(1+\frac{bx}{y}\right)+\frac{b^2-bx}{b^2+y}\log\left(1-\frac{x}{b}\right)\right)\right),$$

$$\leqslant \exp\left(-ng(b,y)x^2\right), \tag{3.30}$$

where

$$g(b,y) = \begin{cases} \dfrac{b^2}{(b^4-y^2)}\log\left(\dfrac{b^2}{y}\right) & \text{if } y < b^2, \\ \dfrac{1}{2y} & \text{if } y \geqslant b^2. \end{cases}$$

Consequently, for any x in $[0,b]$ and for any positive y,

$$\mathbb{P}(M_n \geqslant nx, \langle M \rangle_n \leqslant ny) \leqslant \exp\left(-\frac{nx^2}{2(y+bx/3)}\right). \tag{3.31}$$

Remark 3.11. Note that $M_n \leqslant nb$ almost surely, which implies that $\mathbb{P}(M_n > nb) = 0$. In addition, inequality (3.31) improves the celebrated Freedman's inequality [17] given, for any x in $[0,b]$ and for any positive y, by

$$\mathbb{P}(M_n \geqslant nx, \langle M \rangle_n \leqslant ny) \leqslant \exp\left(-\frac{nx^2}{2(y+bx)}\right). \tag{3.32}$$

We refer the reader to Pinelis [20] for some extension of Freedman's inequality to Banach space valued martingales.

We shall only prove Theorem 3.10 in the special case $b = 1$, inasmuch as the general case follows by dividing the initial martingale by b. The proof of Theorem 3.10 relies on the following lemma where we make use once again of the function L_v defined, for any positive v and for any positive t, by

$$L_v(t) = \log(ve^t + e^{-vt}) - \log(1+v) \tag{3.33}$$

Lemma 3.12. *Let (M_n) be a square integrable martingale such that $M_0 = 0$. Assume that, for all $1 \leqslant k \leqslant n$, $\Delta M_k \leqslant 1$ a.s. For any positive t and for all $n \geqslant 0$, denote*

$$W_n(t) = \exp\left(tM_n - \sum_{k=1}^{n} L_{V_k}(t)\right)$$

with $W_0(t) = 1$. Then, $(W_n(t))$ is a positive supermartingale such that $\mathbb{E}[W_n(t)] \leqslant 1$.

Proof. For any positive t and for all $n \geqslant 1$, we clearly have

$$W_n(t) = W_{n-1}(t)\exp(t\Delta M_n - L_{V_n}(t)).$$

We deduce from Lemma 2.23 that, for any positive t,

$$\mathbb{E}\big[\exp(t\Delta M_n)|\mathscr{F}_{n-1}\big] \leqslant \exp(L_{V_n}(t)) \qquad \text{a.s.}$$

Consequently, for any positive t,

$$\mathbb{E}[W_n(t)|\mathscr{F}_{n-1}] = W_{n-1}(t)\mathbb{E}\big[\exp(t\Delta M_n)|\mathscr{F}_{n-1}\big]\exp(-L_{V_n}(t)) \leqslant W_{n-1}(t).$$

Hence, $(W_n(t))$ is a positive supermartingale. By taking the expectation on both sides of the above inequality, we obtain that for all $n \geqslant 1$, $\mathbb{E}[W_n(t)] \leqslant \mathbb{E}[W_{n-1}(t)]$, which leads to $\mathbb{E}[W_n(t)] \leqslant 1$. $\qquad\qquad\square$

Proof of Theorem 3.10. Denote by f the function defined, for any $v \geqslant 1$, by

$$f(v) = \frac{e^{-vt} + v - 1}{v}.$$

It is not hard to see that, for any positive t, $\log(f(v+1)) = L_v(t) + t$. Moreover, we also have for any $v \geqslant 1$,

$$f(v) = 1 - \int_0^t e^{-vs}\,ds.$$

It implies that f is increasing and strictly concave. Consequently, the function L_v is also increasing and strictly concave with respect to v. It ensures that

$$\frac{1}{n}\sum_{k=1}^{n} L_{V_k}(t) \leqslant L_{\langle M\rangle_n/n}(t). \tag{3.34}$$

We are now in position to prove Theorem 3.10. For any x in $[0,b]$ and for any positive y, denote

$$A_n = \Big\{M_n \geqslant nx, \langle M\rangle_n \leqslant ny\Big\}.$$

By Markov's inequality together with (3.34), we have for any positive t,

$$\mathbb{P}(A_n) \leqslant \mathbb{E}\Big[\exp\big(tM_n - ntx\big)\mathbb{I}_{A_n}\Big],$$

$$\leqslant \mathbb{E}\Big[\exp\big(tM_n - \sum_{k=1}^{n} L_{V_k}(t)\big)\exp\big(\sum_{k=1}^{n} L_{V_k}(t) - ntx\big)\mathbb{I}_{A_n}\Big],$$

$$\leqslant \mathbb{E}\Big[\exp\big(tM_n - \sum_{k=1}^{n} L_{V_k}(t)\big)\exp\big(nL_{\langle M\rangle_n/n}(t) - ntx\big)\mathbb{I}_{A_n}\Big],$$

$$\leqslant \exp\big(n(L_y(t) - tx)\big)\mathbb{E}[W_n(t)],$$

where the last inequality follows from the fact that the function L_v is increasing with respect to v. Therefore, we obtain from Lemma 3.12 that

$$\mathbb{P}(A_n) \leqslant \exp\big(n(L_y(t) - tx)\big).$$

The Legendre-Fenchel transform L_y^* of L_y has been already calculated in (2.55). Hence, the above inequality together with (2.55) and Lemma 2.26 implies (3.30), which achieves the proof of Theorem 3.10 $\qquad\square$

We now deduce improved versions of Freedman's inequalities [17] from Theorem 3.10. The proof, being an immediate consequence of Lemma 2.32, is left to the reader.

Theorem 3.13. *Let* (M_n) *be a square integrable martingale such that* $M_0 = 0$. *Assume that* (M_n) *satisfies* (3.29). *Then, for any x in* $[0,b]$ *and for any positive y,*

$$\mathbb{P}(M_n \geqslant nx, \langle M \rangle_n \leqslant ny) \leqslant \exp\left(-\frac{n}{y}h_w(x)\right) \leqslant \exp\left(-\frac{ny}{b^2}h\left(\frac{bx}{y}\right)\right), \qquad (3.35)$$

where $w = (b/y) - (1/b)$ *and the functions h and* h_w *are given by* (2.60) *and* (2.61), *respectively. Furthermore, for any x in* $[0,b]$ *and for any positive y,*

$$\mathbb{P}(M_n \geqslant nx, \langle M \rangle_n \leqslant ny) \leqslant \exp\left(-\frac{nx^2}{2(y + bx/3)(1 - x/(3b))}\right). \qquad (3.36)$$

3.3 Bernstein's inequality

Bernstein [4] extended his exponential inequalities for sums to martingale differences. More precisely, he obtained an extension of inequality (2.4) under the condition that for integer $p \geqslant 3$ and for all $1 \leqslant k \leqslant n$,

$$\mathbb{E}\left[|\Delta M_k|^p | \mathscr{F}_{k-1}\right] \leqslant \frac{p! c^{p-2}}{2} V_k \qquad \text{a.s.} \qquad (3.37)$$

where (V_n) is the sequence given by (3.2). In this subsection, we will focus our attention on the following improvement of Bernstein's inequality [4].

Theorem 3.14. *Let* (M_n) *be a square integrable martingale such that* $M_0 = 0$. *Assume that there exists a positive constant c such that, for any integer* $p \geqslant 3$ *and for all* $1 \leqslant k \leqslant n$,

$$\mathbb{E}\left[(\max(0, \Delta M_k))^p | \mathscr{F}_{k-1}\right] \leqslant \frac{p! c^{p-2}}{2} V_k \qquad \text{a.s.} \qquad (3.38)$$

Then, for any positive x and for any positive y,

$$\mathbb{P}(M_n \geqslant nx, \langle M \rangle_n \leqslant ny) \leqslant \left(1 + \frac{x^2}{2(y + cx)}\right)^n \exp\left(-\frac{nx^2}{y + cx}\right) \qquad (3.39)$$

$$\leqslant \exp\left(-\frac{nx^2}{2(y + cx)}\right). \qquad (3.40)$$

In addition, we also have, for any positive x and for any positive y,

$$\mathbb{P}(M_n \geqslant nx, \langle M \rangle_n \leqslant ny) \leqslant \exp\left(-\frac{nx^2}{y + cx + \sqrt{y(y + 2cx)}}\right) \tag{3.41}$$

and

$$\mathbb{P}\left(M_n > n(cx + \sqrt{2xy}), \langle M \rangle_n \leqslant ny\right) \leqslant \exp(-nx). \tag{3.42}$$

The proof of Theorem 3.14 relies on the lemma below where the function Λ_v is defined, for any positive v and for any positive t such that $0 < tc < 1$, by

$$\Lambda_v(t) = \log\left(1 + \frac{vt^2}{2(1 - tc)}\right). \tag{3.43}$$

Lemma 3.15. *Let (M_n) be a square integrable martingale such that $M_0 = 0$, satisfying (3.38). For any positive t such that $0 < tc < 1$, and for all $n \geqslant 0$, denote*

$$W_n(t) = \exp\left(tM_n - \sum_{k=1}^{n} \Lambda_{V_k}(t)\right)$$

with $W_0(t) = 1$. Then, $(W_n(t))$ is a positive supermartingale such that $\mathbb{E}[W_n(t)] \leqslant 1$.

Proof. For any positive t such that $0 < tc < 1$, and for all $n \geqslant 1$, we clearly have

$$W_n(t) = W_{n-1}(t)\exp(t\Delta M_n - \Lambda_{V_n}(t)).$$

Proceeding exactly as in the proof of (2.13), we obtain that for all $1 \leqslant k \leqslant n$,

$$\mathbb{E}\left[\exp(t\Delta M_k)|\mathscr{F}_{k-1}\right] \leqslant 1 + \frac{t^2 V_k}{2} + \sum_{p=3}^{\infty} \frac{t^p \mathbb{E}\left[(\max(0, \Delta M_k))^p | \mathscr{F}_{k-1}\right]}{p!} \qquad \text{a.s.}$$

Hence, under assumption (3.38), we find that for any positive t such that $0 < tc < 1$, and for all $1 \leqslant k \leqslant n$,

$$\mathbb{E}\left[\exp(t\Delta M_k)|\mathscr{F}_{k-1}\right] \leqslant 1 + \frac{V_k t^2}{2(1 - tc)} = \exp\left(\Lambda_{V_k}(t)\right) \qquad \text{a.s.}$$

Consequently, for any positive t such that $0 < tc < 1$,

$$\mathbb{E}[W_n(t)|\mathscr{F}_{n-1}] = W_{n-1}(t)\mathbb{E}\left[\exp(t\Delta M_n)|\mathscr{F}_{n-1}\right]\exp(-\Lambda_{V_n}(t)) \leqslant W_{n-1}(t).$$

Finally, $(W_n(t))$ is a positive supermartingale such that, for all $n \geqslant 1$, $\mathbb{E}[W_n(t)] \leqslant 1$, which achieves the proof of Lemma 3.15. $\qquad\square$

Proof of Theorem 3.14. We follow the same lines as in the proof of Theorem 3.10. First of all, one can observe that the function Λ_v is increasing and strictly concave with respect to v. It implies that

$$\frac{1}{n}\sum_{k=1}^{n}\Lambda_{V_k}(t) \leqslant \Lambda_{\langle M\rangle_n/n}(t).$$ (3.44)

For any positive x and for any positive y, denote

$$A_n = \left\{M_n \geqslant nx, \langle M\rangle_n \leqslant ny\right\}.$$

By Markov's inequality together with (3.44), we have for any positive t such that $0 < tc < 1$,

$$\begin{aligned}
\mathbb{P}(A_n) &\leqslant \mathbb{E}\left[\exp\left(tM_n - ntx\right)\mathrm{I}_{A_n}\right], \\
&\leqslant \mathbb{E}\left[\exp\left(tM_n - \sum_{k=1}^{n}\Lambda_{V_k}(t)\right)\exp\left(\sum_{k=1}^{n}\Lambda_{V_k}(t) - ntx\right)\mathrm{I}_{A_n}\right], \\
&\leqslant \mathbb{E}\left[\exp\left(tM_n - \sum_{k=1}^{n}\Lambda_{V_k}(t)\right)\exp\left(n\Lambda_{\langle M\rangle_n/n}(t) - ntx\right)\mathrm{I}_{A_n}\right], \\
&\leqslant \exp\left(n\left(\Lambda_y(t) - tx\right)\right).
\end{aligned}$$ (3.45)

Hereafter, if we choose $t(x) = x/(y + cx)$, we already saw in (2.15) that the Legendre-Fenchel transform Λ_y^* of Λ_y satisfies

$$\Lambda_y^*(x) \geqslant xt(x) - \Lambda_y(t(x)) = \frac{x^2}{y+cx} - \log\left(1 + \frac{x^2}{2(y+cx)}\right).$$ (3.46)

Consequently, (3.39) and (3.40) immediately follows from (3.45) and (3.46). Furthermore, as (3.41) and (3.42) are equivalent, it only remains to prove (3.42). We deduce from the elementary inequality $\log(1 + x) \leqslant x$ that for any positive t such that $0 < tc < 1$, $\Lambda_y(t) \leqslant \ell_y(t)$ where

$$\ell_y(t) = \frac{yt^2}{2(1-tc)}.$$

Hence, we obtain from (3.45) that for any positive x and for any positive y,

$$\mathbb{P}(M_n \geqslant nx, \langle M\rangle_n \leqslant ny) \leqslant \exp(n(\ell_y(t) - tx))$$

which ensures that

$$\mathbb{P}\left(M_n \geqslant n\inf_{tc\in]0,1[}\left(\frac{\ell_y(t)+x}{t}\right), \langle M\rangle_n \leqslant ny\right) \leqslant \exp(-nx).$$ (3.47)

However, we already saw that

$$\inf_{tc\in]0,1[}\left(\frac{\ell_y(t)+x}{t}\right) = \inf_{tc\in]0,1[}\left(\frac{yt}{2(1-tc)} + \frac{x}{t}\right) = cx + \sqrt{2xy}$$ (3.48)

where the infimum is given by the optimal value $t = \sqrt{2x}/(\sqrt{y} + c\sqrt{2x})$. Finally, we obtain (3.42) from (3.47) and (3.48), which completes the proof of Theorem 3.14. $\qquad\square$

3.4 De la Pena's inequalities

In order to avoid any boundedness or moment assumption, De la Peña [6] proposes new exponential inequalities in the particular case where (M_n) is a conditionally symmetric martingale. It involves its total quadratic variation

$$[M]_n = \sum_{k=1}^{n} \Delta M_k^2.$$

3.4.1 Conditionally symmetric martingales

Definition 3.16. Let (M_n) be a martingale adapted to a filtration $\mathbb{F} = (\mathscr{F}_n)$. We shall say that (M_n) is conditionally symmetric if, for all $n \geqslant 1$, the distribution of its increments ΔM_n given $\mathscr{F}_{n-1}$ is symmetric.

Theorem 3.17. *Let (M_n) be a square integrable and conditionally symmetric martingale such that $M_0 = 0$. Then, for any positive x and for any positive y,*

$$\mathbb{P}(M_n \geqslant x, [M]_n \leqslant y) \leqslant \exp\left(-\frac{x^2}{2y}\right). \tag{3.49}$$

For self-normalized martingales, the result is as follows.

Theorem 3.18. *Let (M_n) be a square integrable and conditionally symmetric martingale, such that $M_0 = 0$. Then, for any positive x and y, and for all $a \geqslant 0$, $b > 0$*

$$\mathbb{P}\left(\frac{M_n}{a+b[M]_n} \geqslant x\right) \leqslant \sqrt{\mathbb{E}\left[\exp\left(-\frac{x^2}{2}\left(2ab + b^2[M]_n\right)\right)\right]}, \tag{3.50}$$

$$\mathbb{P}\left(\frac{M_n}{a+b[M]_n} \geqslant x, [M]_n \geqslant y\right) \leqslant \exp\left(-\frac{x^2}{2}\left(2ab + b^2y\right)\right). \tag{3.51}$$

Remark 3.19. One can find in [7, 11, 14] some interesting extensions of the above inequalities. We also refer the reader to the recent survey of De la Peña, Klass, and Lai [8]. However, the conditionally symmetric assumption is still required for (3.49). By a careful reading of [6], one can realize that (3.49) is a two-sided exponential inequality. More precisely, if (M_n) is a square integrable and conditionally symmetric martingale such that $M_0 = 0$, then for any positive x and y,

$$\mathbb{P}(|M_n| \geqslant x, [M]_n \leqslant y) \leqslant 2\exp\left(-\frac{x^2}{2y}\right). \tag{3.52}$$

Finally, the proofs of Theorems 3.17 and 3.18 are omitted as we shall see in Section 3.6 how to extend those results relaxing the assumption that the martingale (M_n) is conditionally symmetric.

3.4.2 The missing factors

We shall now establish concentration inequalities for self-normalized martingales in connection with the central limit theorem. Our goal is to provide missing factors in exponential inequalities for self-normalized martingales with upper bounds independent of $[M]_n$ or $\langle M \rangle_n$. We refer the reader to the interesting monograph of De la Peña, Lai, and Shao [10] for a comprehensive review of theoretical results on self-normalized processes with statistical applications. First of all, De la Peña and Pang [9] recently improve a result due to De la Peña, Klass, and Lai [7], as follows.

Theorem 3.20. *Let (M_n) be a square integrable and conditionally symmetric martingale such that $M_0 = 0$. In addition, assume that $\mathbb{E}[|M_n|^p] < \infty$ for some $p \geqslant 2$. Then, for any positive x,*

$$\mathbb{P}\left(\frac{|M_n|}{\sqrt{[M]_n + (\mathbb{E}[|M_n|^p])^{2/p}}} \geqslant x\sqrt{\frac{2q-1}{q}}\right) \leqslant C_q x^{-q/(2q-1)}\exp\left(-\frac{x^2}{2}\right) \tag{3.53}$$

where $q = p/(p-1)$ is the Hölder conjugate exponent of p and

$$C_q = \left(\frac{q}{2q-1}\right)^{q/2(2q-1)}.$$

In particular, for $p = 2$, we have for any positive x,

$$\mathbb{P}\left(\frac{|M_n|}{\sqrt{[M]_n + \mathbb{E}[M_n^2]}} \geqslant x\sqrt{\frac{3}{2}}\right) \leqslant \left(\frac{2}{3}\right)^{1/3} x^{-2/3}\exp\left(-\frac{x^2}{2}\right). \tag{3.54}$$

The proof of Theorem 3.20 relies on the following keystone lemma.

Lemma 3.21. *For a symmetric random variable X and for any real t,*

$$L(t) = \mathbb{E}\left[\exp\left(tX - \frac{t^2}{2}X^2\right)\right] \leqslant 1. \tag{3.55}$$

Proof. Let F be the cumulative distribution function of the random variable X. Denote by f the function defined, for any real x, by

$$f(x) = \exp\left(x - \frac{x^2}{2}\right).$$

As X is symmetric, we have for all real t,

$$L(t) = \int_{\mathbb{R}} f(tx)\,dF(x) = \int_0^{+\infty} (f(tx) + f(-tx))\,dF(x),$$

$$= 2\int_0^{+\infty} \exp(-t^2 x^2/2)\cosh(tx)\,dF(x) \leqslant 1$$

by the well-known inequality $\cosh(x) \leqslant \exp(x^2/2)$. $\qquad\qquad\square$

Proof of Theorem 3.20. First of all, for any real t and for all $n \geqslant 0$, denote

$$W_n(t) = \exp\left(tM_n - \frac{t^2}{2}[M]_n\right)$$

with $W_0(t) = 1$. We claim that for any real t, $(W_n(t))$ is a positive supermartingale such that $\mathbb{E}[W_n(t)] \leqslant 1$. As a matter of fact, we have for any real t and for all $n \geqslant 0$,

$$W_n(t) = W_{n-1}(t)\exp\left(t\Delta M_n - \frac{t^2}{2}\Delta M_n^2\right).$$

However, it follows from Definition 3.16 and Lemma 3.21 that for any real t,

$$\mathbb{E}\left[\exp\left(t\Delta M_n - \frac{t^2}{2}\Delta M_n^2\right)\Big|\mathscr{F}_{n-1}\right] \leqslant 1 \qquad \text{a.s.}$$

which implies that $\mathbb{E}[W_n(t)|\mathscr{F}_{n-1}] \leqslant W_{n-1}(t)$ a.s. Hence, $(W_n(t))$ is a positive supermartingale such that, for all $n \geqslant 1$, $\mathbb{E}[W_n(t)] \leqslant 1$. For any positive C, let X be a random variable with Gaussian $\mathscr{N}(0, 1/C)$ distribution. As

$$\sqrt{\frac{C}{2\pi}} \int_{\mathbb{R}} \exp\left(-\frac{Cx^2}{2}\right)dx = 1,$$

we clearly have

$$\sqrt{\frac{C}{2\pi}} \int_{\mathbb{R}} \exp\left(-\frac{Cx^2}{2}\right)\mathbb{E}\left[W_n(x)\right]dx \leqslant 1. \qquad (3.56)$$

However, if $D_n = [M]_n + C$, it follows from Fubini's theorem that

$$\sqrt{\frac{C}{2\pi}} \int_{\mathbb{R}} \exp\left(-\frac{Cx^2}{2}\right)\mathbb{E}\left[W_n(x)\right]dx = \mathbb{E}\left[\sqrt{\frac{C}{2\pi}} \int_{\mathbb{R}} \exp\left(-\frac{D_n x^2}{2} + xM_n\right)dx\right],$$

$$= \mathbb{E}\left[\sqrt{\frac{C}{2\pi}}\exp\left(\frac{M_n^2}{2D_n}\right)\int_{\mathbb{R}} \exp\left(-\frac{D_n}{2}\left(x - \frac{M_n}{D_n}\right)^2\right)dx\right],$$

$$= \mathbb{E}\left[\sqrt{\frac{C}{D_n}}\exp\left(\frac{M_n^2}{2D_n}\right)\right]. \qquad (3.57)$$

Consequently, we immediately deduce from (3.56) and (3.57) that

$$\mathbb{E}\left[\sqrt{\frac{C}{D_n}}\exp\left(\frac{M_n^2}{2D_n}\right)\right] \leqslant 1. \tag{3.58}$$

We are now in position to prove Theorem 3.20. For any positive x, let

$$A_n = \left\{|M_n| \geqslant x\sqrt{D_n}\right\}.$$

It follows from Markov's inequality that

$$\mathbb{P}(A_n) = \mathbb{P}\left(\frac{M_n^2}{D_n} \geqslant x^2\right) = \mathbb{P}\left(\left(\frac{M_n^2}{D_n}\right)^{1/4}\exp\left(\frac{M_n^2}{4D_n}\right) \geqslant \sqrt{x}\exp\left(\frac{x^2}{4}\right)\right),$$
$$\leqslant x^{-1/2}\exp\left(-\frac{x^2}{4}\right)\mathbb{E}\left[\left(\frac{M_n^2}{D_n}\right)^{1/4}\exp\left(\frac{M_n^2}{4D_n}\right)\mathrm{I}_{A_n}\right]. \tag{3.59}$$

Moreover, we deduce from Cauchy-Schwarz's inequality together with (3.58) that

$$\mathbb{E}\left[\left(\frac{M_n^2}{D_n}\right)^{1/4}\exp\left(\frac{M_n^2}{4D_n}\right)\mathrm{I}_{A_n}\right] = \mathbb{E}\left[\left(\frac{C}{D_n}\right)^{1/4}\exp\left(\frac{M_n^2}{4D_n}\right)\left(\frac{M_n^2}{C}\right)^{1/4}\mathrm{I}_{A_n}\right],$$
$$\leqslant \left(\mathbb{E}\left[\sqrt{\frac{C}{D_n}}\exp\left(\frac{M_n^2}{2D_n}\right)\right]\right)^{1/2}\left(\mathbb{E}\left[\sqrt{\frac{M_n^2}{C}}\mathrm{I}_{A_n}\right]\right)^{1/2},$$
$$\leqslant \left(\mathbb{E}\left[\sqrt{\frac{M_n^2}{C}}\mathrm{I}_{A_n}\right]\right)^{1/2}. \tag{3.60}$$

Hereafter, if we choose $C = \left(\mathbb{E}[|M_n|^p]\right)^{2/p}$, we obtain from Hölder's inequality that

$$\mathbb{E}\left[\sqrt{\frac{M_n^2}{C}}\mathrm{I}_{A_n}\right] \leqslant \frac{1}{\sqrt{C}}\left(\mathbb{E}[|M_n|^p]\right)^{1/p}\left(\mathbb{P}(A_n)\right)^{1/q} \leqslant \left(\mathbb{P}(A_n)\right)^{1/q} \tag{3.61}$$

where $q = p/(p-1)$ is the Hölder conjugate exponent of p. Therefore, we find from (3.59), (3.60), and (3.61) that

$$\mathbb{P}(A_n) \leqslant x^{-1/2}\exp\left(-\frac{x^2}{4}\right)\left(\mathbb{P}(A_n)\right)^{1/q}$$

which leads to

$$\mathbb{P}(A_n) \leqslant x^{-q/(2q-1)}\exp\left(-\frac{qx^2}{2(2q-1)}\right). \tag{3.62}$$

Finally, we obtain inequality (3.53) replacing x by $x\sqrt{2q-1}/\sqrt{q}$ into (3.62), which achieves the proof of Theorem 3.20. $\qquad\qquad\square$

Without any assumption on (M_n), the second result involves both the total quadratic variation $[M]_n$ and the increasing process $\langle M \rangle_n$.

Theorem 3.22. *Let (M_n) be a square integrable martingale such that $M_0 = 0$. Then, for any positive x,*

$$\mathbb{P}\left(\frac{|M_n|}{\sqrt{[M]_n + \langle M \rangle_n + \mathbb{E}[M_n^2]}} \geq x\sqrt{\frac{3}{2}} \right) \leq \left(\frac{2}{3}\right)^{1/3} x^{-2/3} \exp\left(-\frac{x^2}{2}\right). \tag{3.63}$$

This result can be slightly improved as follows.

Theorem 3.23. *Let (M_n) be a square integrable martingale such that $M_0 = 0$. Then, for any positive x,*

$$\mathbb{P}\left(\frac{|M_n|}{\sqrt{[M]_n + 2\langle M \rangle_n + 3\mathbb{E}[M_n^2]}} \geq \frac{x}{\sqrt{2}} \right) \leq \left(\frac{2}{3}\right)^{1/3} x^{-2/3} \exp\left(-\frac{x^2}{2}\right). \tag{3.64}$$

Proof. For any real t and for all $n \geq 0$, denote

$$V_n(t) = \exp\left(tM_n - \frac{t^2}{6}[M]_n - \frac{t^2}{3}\langle M \rangle_n \right).$$

We will see in Lemma 3.34 below that, without any assumption on (M_n), $(V_n(t))$ is a supermartingale such that $\mathbb{E}[V_n(t)] \leq 1$. Hence, via the same lines as in the proof of Theorem 3.20 in the particular case $p = 2$, we obtain that for any positive x,

$$\mathbb{P}\left(\frac{|M_n|}{\sqrt{D_n}} \geq x\sqrt{\frac{3}{2}} \right) \leq \left(\frac{2}{3}\right)^{1/3} x^{-2/3} \exp\left(-\frac{x^2}{2}\right)$$

where

$$D_n = \mathbb{E}[M_n^2] + \frac{1}{3}[M]_n + \frac{2}{3}\langle M \rangle_n$$

which immediately leads to (3.64). □

3.5 Gaussian martingales

Our goal is now to establish exponential inequalities for (M_n) similar to (3.49), (3.50), and (3.51), replacing the total quadratic variation $[M]_n$ by the increasing process $\langle M \rangle_n$. For that purpose, it is necessary to assume that the martingale (M_n) is Gaussian.

Definition 3.24. Let (M_n) be a martingale adapted to a filtration $\mathbb{F} = (\mathscr{F}_n)$. We shall say that (M_n) is Gaussian if, for all $n \geq 1$, the distribution of its increments ΔM_n given $\mathscr{F}_{n-1}$ is $\mathscr{N}(0, \Delta\langle M \rangle_n)$.

Theorem 3.25. *Let (M_n) be a square integrable Gaussian martingale such that $M_0 = 0$. Then, all the results of Theorems 3.17 and 3.18 are true, replacing $[M]_n$ by $\langle M \rangle_n$ everywhere. For example, for any positive x and y,*

$$\mathbb{P}(M_n \geqslant x, \langle M \rangle_n \leqslant y) \leqslant \exp\left(-\frac{x^2}{2y}\right). \tag{3.65}$$

Moreover, for any positive x, and for all $a \geqslant 0$ and $b > 0$,

$$\mathbb{P}\left(\frac{M_n}{a + b\langle M \rangle_n} \geqslant x\right) \leqslant \inf_{p>1}\left(\mathbb{E}\left[\exp\left(-\frac{(p-1)x^2}{2}\left(2ab + b^2\langle M \rangle_n\right)\right)\right]\right)^{1/p}. \tag{3.66}$$

Proof. First of all, for any real t and for all $n \geqslant 0$, let

$$W_n(t) = \exp\left(tM_n - \frac{t^2}{2}\langle M \rangle_n\right).$$

Since (M_n) is a Gaussian martingale, $(W_n(t))$ is a positive martingale such that $\mathbb{E}[W_n(t)] = 1$. For any positive x and for any positive y, denote

$$A_n = \left\{M_n \geqslant x, \langle M \rangle_n \leqslant y\right\}.$$

It follows from Markov's inequality together with Cauchy-Schwarz's inequality that

$$\begin{aligned}
\mathbb{P}(A_n) &\leqslant \mathbb{E}\left[\exp\left(\frac{t}{2}M_n - \frac{tx}{2}\right)\mathrm{I}_{A_n}\right], \\
&\leqslant \mathbb{E}\left[\sqrt{W_n(t)}\exp\left(\frac{t^2}{4}\langle M \rangle_n - \frac{tx}{2}\right)\mathrm{I}_{A_n}\right], \\
&\leqslant \exp\left(\frac{t^2 y}{4} - \frac{tx}{2}\right)\sqrt{\mathbb{P}(A_n)} \tag{3.67}
\end{aligned}$$

where the last inequality is due to the fact that $\mathbb{E}[W_n(t)] \leqslant 1$. Therefore, dividing both sides of (3.67) by $\sqrt{\mathbb{P}(A_n)}$ and choosing the optimal value $t = x/y$, we obtain that

$$\mathbb{P}(A_n) \leqslant \exp\left(-\frac{x^2}{2y}\right),$$

which is exactly (3.65). We shall continue the proof of Theorem 3.25 in the particular case $a = 0$ and $b = 1$, inasmuch as the proof of (3.66) in the general case follows exactly the same lines. For any positive x, let

$$B_n = \left\{M_n \geqslant x\langle M \rangle_n\right\}.$$

We deduce from Markov's inequality together with Holder's inequality that, for any positive t and for all $q > 1$,

$$\mathbb{P}(B_n) \leqslant \mathbb{E}\left[\exp\left(\frac{t}{q}M_n - \frac{tx}{q}\langle M\rangle_n\right)\mathrm{I}_{B_n}\right],$$

$$\leqslant \mathbb{E}\left[(W_n(t))^{1/q}\exp\left(\frac{t}{2q}(t-2x)\langle M\rangle_n\right)\mathrm{I}_{B_n}\right],$$

$$\leqslant \left(\mathbb{E}\left[\exp\left(\frac{tp}{2q}(t-2x)\langle M\rangle_n\right)\right]\right)^{1/p} \tag{3.68}$$

where p and q are Hölder conjugate exponents and $\mathbb{E}[W_n(t)] = 1$. Consequently, we deduce from (3.68) with $t = x$ and the elementary fact that $p/q = p - 1$, that

$$\mathbb{P}(B_n) \leqslant \inf_{p>1}\left(\mathbb{E}\left[\exp\left(-\frac{(p-1)x^2}{2}\langle M\rangle_n\right)\right]\right)^{1/p}$$

which immediately implies (3.66). In the particular case $p = 2$, we obtain that

$$\mathbb{P}(B_n) \leqslant \sqrt{\mathbb{E}\left[\exp\left(-\frac{x^2}{2}\langle M\rangle_n\right)\right]}.$$

Finally, for any positive x and for any positive y, denote

$$C_n = \left\{M_n \geqslant x\langle M\rangle_n, \langle M\rangle_n \geqslant y\right\}.$$

We clearly have

$$\mathbb{P}(C_n) \leqslant \mathbb{E}\left[\exp\left(xM_n - x^2\langle M\rangle_n\right)\mathrm{I}_{C_n}\right],$$

$$\leqslant \exp\left(-\frac{x^2 y}{2}\right)\mathbb{E}\left[W_n(x)\right],$$

$$\leqslant \exp\left(-\frac{x^2 y}{2}\right)$$

which achieves the proof of Theorem 3.25. $\qquad\qquad\square$

3.6 Always a little further on martingales

We shall now propose an exponential inequality for (M_n) which holds without any assumption. It involves both the total quadratic variation $[M]_n$ and the increasing process $\langle M\rangle_n$.

Theorem 3.26. *Let (M_n) be a square integrable martingale such that $M_0 = 0$. Then, for any positive x and y,*

$$\mathbb{P}\left(M_n \geqslant x, [M]_n + \langle M\rangle_n \leqslant y\right) \leqslant \exp\left(-\frac{x^2}{2y}\right). \tag{3.69}$$

Remark 3.27. We refer the reader to Bercu and Touati [3] for more details concerning this result. Similar result for continuous-time locally square integrable martingale may be found in Barlow, Jacka, and Yor [2].

For self-normalized martingales, the results are as follows.

Theorem 3.28. *Let (M_n) be a square integrable martingale such that $M_0 = 0$. Then, for any positive x and y, and for all $a \geqslant 0$ and $b > 0$,*

$$\mathbb{P}\left(\frac{M_n}{a+b\langle M\rangle_n} \geqslant x, \langle M\rangle_n \geqslant [M]_n + y\right) \leqslant \exp\left(-\frac{x^2}{2}\left(2ab + b^2 y\right)\right). \qquad (3.70)$$

Moreover, we also have

$$\mathbb{P}\left(\frac{M_n}{a+b\langle M\rangle_n} \geqslant x, [M]_n \leqslant y\langle M\rangle_n\right) \qquad (3.71)$$

$$\leqslant \inf_{p>1}\left(\mathbb{E}\left[\exp\left(-\frac{(p-1)x^2}{2(1+y)}\left(2ab + b^2\langle M\rangle_n\right)\right)\right]\right)^{1/p}.$$

Remark 3.29. It is not hard to see that (3.70) and (3.71) also hold exchanging the roles of $\langle M\rangle_n$ and $[M]_n$.

Those exponential inequalities were recently improved by Delyon [11] as follows.

Theorem 3.30. *Let (M_n) be a square integrable martingale such that $M_0 = 0$. Then, for any positive x and y,*

$$\mathbb{P}\left(M_n \geqslant x, [M]_n + 2\langle M\rangle_n \leqslant y\right) \leqslant \exp\left(-\frac{3x^2}{2y}\right). \qquad (3.72)$$

Remark 3.31. For any positive x and y, as

$$\mathbb{P}\left(M_n \geqslant x, [M]_n + \langle M\rangle_n \leqslant y\right) \leqslant \mathbb{P}\left(M_n \geqslant x, [M]_n + 2\langle M\rangle_n \leqslant 2y\right),$$

$$\leqslant \exp\left(-\frac{3x^2}{4y}\right) \leqslant \exp\left(-\frac{x^2}{2y}\right),$$

one can easily see that inequality (3.72) is sharper than (3.69).

Theorem 3.32. *Let (M_n) be a square integrable martingale such that $M_0 = 0$. Then, for any positive x and y, and for all $a \geqslant 0$ and $b > 0$,*

$$\mathbb{P}\left(\frac{M_n}{a+b\langle M\rangle_n} \geqslant x, 2\langle M\rangle_n \geqslant [M]_n + 2y\right) \leqslant \exp\left(-\frac{3x^2}{4}\left(2ab + b^2 y\right)\right). \qquad (3.73)$$

Moreover, we also have

$$\mathbb{P}\left(\frac{M_n}{a+b\langle M\rangle_n} \geqslant x, [M]_n \leqslant 2y\langle M\rangle_n\right) \tag{3.74}$$

$$\leqslant \inf_{p>1}\left(\mathbb{E}\left[\exp\left(-\frac{3(p-1)x^2}{4(1+y)}\left(2ab+b^2\langle M\rangle_n\right)\right)\right]\right)^{1/p}.$$

The proof of Theorem 3.26 relies on the inequality given, for any real x, by

$$f(x) = \exp\left(x - \frac{x^2}{2}\right) \leqslant g(x) = 1+x+\frac{x^2}{2}, \tag{3.75}$$

whereas that of Theorem 3.30 is based on the inequality given, for any real x, by

$$h(x) = \exp\left(x - \frac{x^2}{6}\right) \leqslant \ell(x) = 1+x+\frac{x^2}{3}. \tag{3.76}$$

One can observe that for any real x,

$$f(x) \leqslant h(x) \leqslant \ell(x) \leqslant g(x)$$

which also explain why (3.72) is sharper than (3.69), see also Figure 3.1 below.

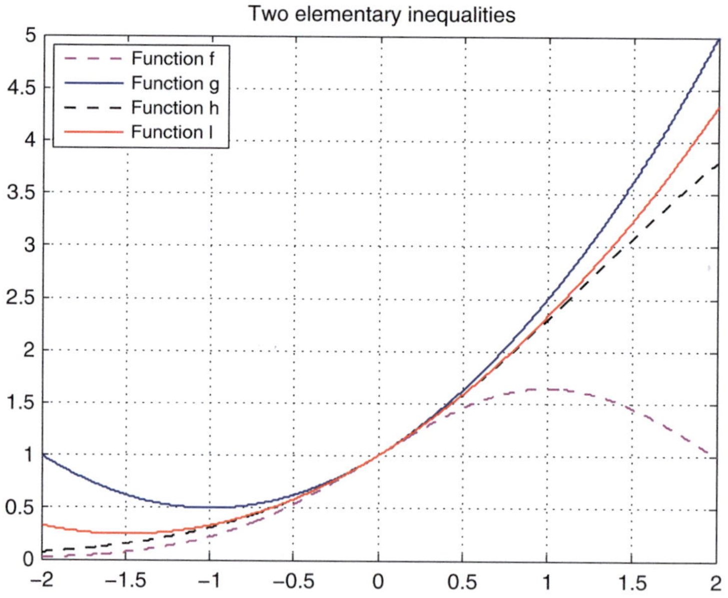

Fig. 3.1 Comparison of the functions f, g, h, ℓ

Lemma 3.33. *Let X be a square integrable random variable with mean zero and variance σ^2. Then, for any real t,*

$$L(t) = \mathbb{E}\left[\exp\left(tX - \frac{t^2}{6}X^2\right)\right] \leqslant 1 + \frac{t^2}{3}\sigma^2. \tag{3.77}$$

Proof. First of all, we prove that inequality (3.76) holds true. For any real x, let

$$\varphi(x) = 1 + x + \frac{x^2}{3} - \exp\left(x - \frac{x^2}{6}\right).$$

We clearly have

$$\varphi'(x) = 1 + \frac{2x}{3} - \left(1 - \frac{x}{3}\right)\exp\left(x - \frac{x^2}{6}\right),$$

$$\varphi''(x) = \frac{2}{3}\left(1 - \left(1 - x + \frac{x^2}{6}\right)\exp\left(x - \frac{x^2}{6}\right)\right).$$

However, for any real x, $\exp(x) \geqslant 1 + x$ which implies that

$$\exp\left(-x + \frac{x^2}{6}\right) \geqslant 1 - x + \frac{x^2}{6}, \qquad 1 \geqslant \left(1 - x + \frac{x^2}{6}\right)\exp\left(x - \frac{x^2}{6}\right).$$

It ensures that for any real x, $\varphi''(x) \geqslant 0$, which means that φ is a convex function. Since $\varphi'(0) = 0$ and $\varphi(0) = 0$, we find that for any real x, $\varphi(x) \geqslant 0$ which is exactly what we wanted to prove. Therefore, we immediately obtain from (3.76) that for any real t,

$$L(t) = \mathbb{E}[h(tX)] \leqslant \mathbb{E}[\ell(tX)] = 1 + \frac{t^2}{3}\sigma^2,$$

which completes the proof of Lemma 3.33. $\qquad\square$

Lemma 3.34. *For any real t and for all $n \geqslant 0$, let*

$$V_n(t) = \exp\left(tM_n - \frac{t^2}{6}[M]_n - \frac{t^2}{3}\langle M \rangle_n\right).$$

Then, $(V_n(t))$ is a positive supermartingale such that $\mathbb{E}[V_n(t)] \leqslant 1$.

Proof. For any real t and for all $n \geqslant 1$, we clearly have

$$V_n(t) = V_{n-1}(t)\exp\left(t\Delta M_n - \frac{t^2}{6}\Delta[M]_n - \frac{t^2}{3}\Delta\langle M \rangle_n\right)$$

where $\Delta M_n = M_n - M_{n-1}$, $\Delta[M]_n = \Delta M_n^2$ and $\Delta\langle M \rangle_n = \mathbb{E}[\Delta M_n^2 | \mathscr{F}_{n-1}]$. Hence, we deduce from Lemma 3.33 that for any real t,

$$\mathbb{E}[V_n(t)|\mathscr{F}_{n-1}] \leqslant V_{n-1}(t)\exp\left(-\frac{t^2}{3}\Delta\langle M \rangle_n\right)\left(1 + \frac{t^2}{3}\Delta\langle M \rangle_n\right),$$
$$\leqslant V_{n-1}(t)$$

via the elementary inequality $1 + x \leqslant \exp(x)$. Consequently, $(V_n(t))$ is a positive supermartingale satisfying $\mathbb{E}[V_n(t)] \leqslant \mathbb{E}[V_{n-1}(t)]$, which leads to $\mathbb{E}[V_n(t)] \leqslant 1$. $\square$

Proof of Theorem 3.30. We are now in position to prove Theorem 3.30. First of all, denote

$$Z_n = \frac{1}{3}[M]_n + \frac{2}{3}\langle M \rangle_n.$$

In addition, for any positive x and y, let

$$A_n = \left\{ M_n \geqslant x, 3Z_n \leqslant y \right\}.$$

By Markov's inequality, we have for any positive t,

$$\mathbb{P}(A_n) \leqslant \mathbb{E}\left[\exp\left(tM_n - tx \right) \mathrm{I}_{A_n} \right],$$
$$\leqslant \mathbb{E}\left[\exp\left(tM_n - \frac{t^2}{2}Z_n \right) \exp\left(\frac{t^2}{2}Z_n - tx \right) \mathrm{I}_{A_n} \right],$$
$$\leqslant \exp\left(\frac{t^2 y}{6} - tx \right) \mathbb{E}[V_n(t)],$$
$$\leqslant \exp\left(\frac{t^2 y}{6} - tx \right).$$

Hence, by taking the optimal value $t = 3x/y$ in the above inequality, we find that

$$\mathbb{P}(A_n) \leqslant \exp\left(-\frac{3x^2}{2y} \right),$$

which is exactly what we wanted to prove. $\square$

Proof of Theorem 3.32. We shall now proceed to the proof of Theorem 3.32. For any positive x and y, let

$$B_n = \left\{ M_n \geqslant x(a + b\langle M \rangle_n), 2\langle M \rangle_n \geqslant [M]_n + 2y \right\}.$$

We have for any positive t,

$$\mathbb{P}(B_n) \leqslant \mathbb{E}\left[\exp\left(\frac{t}{2}M_n - \frac{tax}{2} - \frac{tbx}{2}\langle M \rangle_n \right) \mathrm{I}_{B_n} \right],$$
$$\leqslant \exp\left(-\frac{tax}{2} \right) \mathbb{E}\left[\exp\left(\frac{t}{2}M_n - \frac{t^2}{4}Z_n \right) \exp\left(\frac{t}{6}(t - 3bx)\langle M \rangle_n + \frac{t^2}{12}[M]_n \right) \mathrm{I}_{B_n} \right].$$

Consequently, we obtain from the above inequality with the optimal choice $t = 3bx/2$, together with Cauchy-Schwarz's inequality, that

$$\mathbb{P}(B_n) \leqslant \exp\left(-\frac{3x^2}{8}(2ab+b^2y)\right)\mathbb{E}[\sqrt{V_n(t)}\mathrm{I}_{B_n}],$$

$$\leqslant \exp\left(-\frac{3x^2}{8}(2ab+b^2y)\right)\sqrt{\mathbb{P}(B_n)}. \tag{3.78}$$

Therefore, if we divide both sides of (3.78) by $\sqrt{\mathbb{P}(B_n)}$, we find that

$$\mathbb{P}(B_n) \leqslant \exp\left(-\frac{3x^2}{4}(2ab+b^2y)\right),$$

which clearly implies (3.73). Furthermore, for any positive x and y, let

$$C_n = \left\{M_n \geqslant x(a+b\langle M\rangle_n), [M]_n \leqslant 2y\langle M\rangle_n\right\}.$$

By Holder's inequality, we have for any positive t and $q > 1$,

$$\mathbb{P}(C_n) \leqslant \mathbb{E}\left[\exp\left(\frac{t}{q}M_n - \frac{tax}{q} - \frac{tbx}{q}\langle M\rangle_n\right)\mathrm{I}_{C_n}\right],$$

$$\leqslant \exp\left(-\frac{tax}{q}\right)\mathbb{E}\left[\exp\left(\frac{t}{q}M_n-\frac{t^2}{2q}Z_n\right)\exp\left(\frac{t}{3q}(t(1+y)-3bx)\langle M\rangle_n\right)\mathrm{I}_{C_n}\right],$$

$$\leqslant \exp\left(-\frac{tax}{q}\right)\mathbb{E}\left[(V_n(t))^{1/q}\exp\left(\frac{t}{3q}(t(1+y)-3bx)\langle M\rangle_n\right)\right],$$

$$\leqslant \exp\left(-\frac{tax}{q}\right)\left(\mathbb{E}\left[\exp\left(\frac{tp}{3q}(t(1+y)-3bx)\langle M\rangle_n\right)\right]\right)^{1/p} \tag{3.79}$$

where p and q are Hölder conjugate exponents. Consequently, as $p/q = p-1$, we can deduce from (3.79) with the optimal value $t = 3bx/(2(1+y))$ that

$$\mathbb{P}(C_n) \leqslant \inf_{p>1}\left(\mathbb{E}\left[\exp\left(-\frac{3(p-1)x^2}{4(1+y)}\left(2ab+b^2\langle M\rangle_n\right)\right)\right]\right)^{1/p}$$

which completes the proof of Theorem 3.32. $\qquad\square$

3.7 Martingales heavy on left or right

We shall now introduce a new notion of martingales heavy on left or right [3] which allows us to improve De la Peña's inequalities given in Section 3.4.

Definition 3.35. We shall say that an integrable random variable X is heavy on left if $\mathbb{E}[X] = 0$ and, for any positive a, $\mathbb{E}[T_a(X)] \leqslant 0$ where T_a is the truncation function given, for all real x, by

$$T_a(x) = \begin{cases} a & \text{if} & x \geqslant a, \\ x & \text{if} & -a \leqslant x \leqslant a, \\ -a & \text{if} & x \leqslant -a. \end{cases}$$

Moreover, we shall say that X is heavy on right if $-X$ is heavy on left.

Remark 3.36. Let F be the cumulative distribution function associated with X. Standard calculation leads to $\mathbb{E}[T_a(X)] = -H(a)$ where H stands for the function defined, for any positive a, by

$$H(a) = \int_0^a F(-x) - (1 - F(x)) \, dx.$$

Consequently, X is heavy on left if $\mathbb{E}[X] = 0$ and, for any positive a, $H(a) \geqslant 0$. Moreover, H is equal to zero at infinity as

$$\lim_{a \to \infty} H(a) = -\mathbb{E}[X] = 0.$$

Furthermore, on can observe that a random variable X is symmetric if and only if X is heavy on left and on right.

We shall now provide several examples of random variables heavy on left. We wish to point out that most of all positive random variables centered around their mean are heavy on left. As a matter of fact, let Y be a positive integrable random variable with mean m and denote

$$X = Y - m.$$

Discrete random variables.

1) If Y has a Bernoulli $\mathscr{B}(p)$ distribution with parameter $0 < p < 1$, then X is heavy on left, heavy on right, or symmetric if $p < 1/2$, $p > 1/2$, or $p = 1/2$, respectively.
2) If Y has a Geometric $\mathscr{G}(p)$ distribution with parameter $0 < p < 1$, then X is always heavy on left.
3) If Y has a Poisson $\mathscr{P}(\lambda)$ distribution where the parameter λ is a positive integer, then X is heavy on left.

Continuous random variables.

1) If Y has an exponential $\mathscr{E}(\lambda)$ distribution with parameter $\lambda > 0$, then X is always heavy on left.
2) If Y has a Gamma $\Gamma(a, \lambda)$ distribution with parameters $a, \lambda > 0$, then X is always heavy on left.
3) If Y has a Pareto distribution with parameters $a, \lambda > 0$, which means that $Y = a \exp(Z)$ where Z has an exponential $\mathscr{E}(\lambda)$ distribution, then X is always heavy on left.

Definition 3.37. Let (M_n) be a locally square integrable martingale adapted to a filtration $\mathbb{F} = (\mathscr{F}_n)$. We shall say that (M_n) is heavy on left if all its increments are conditionally heavy on left. In other words, for all $n \geqslant 1$ and for any positive a, $\mathbb{E}[T_a(\Delta M_n)|\mathscr{F}_{n-1}] \leqslant 0$. Moreover, (M_n) is heavy on right if $(-M_n)$ is heavy on left, whereas (M_n) is conditionally symmetric if and only if (M_n) is heavy on left and on right.

Our strategy is quite simple. If (M_n) is heavy on left, one can establish an exponential inequality on the right side, while if (M_n) is heavy on right, one can prove an exponential inequality on the left side. Of course, if (M_n) is conditionally symmetric, two-sided exponential inequalities are straightforward.

Theorem 3.38. *Let* (M_n) *be a square integrable martingale heavy on left such that* $M_0 = 0$. *Then, for any positive* x *and* y,

$$\mathbb{P}\Big(M_n \geqslant x, [M]_n \leqslant y\Big) \leqslant \exp\Big(-\frac{x^2}{2y}\Big). \tag{3.80}$$

For self-normalized martingales, the results are as follows:

Theorem 3.39. *Let* (M_n) *be a square integrable martingale heavy on left such that* $M_0 = 0$. *Then, for any positive* x *and* y, *and for all* $a \geqslant 0$ *and* $b > 0$,

$$\mathbb{P}\left(\frac{M_n}{a+b[M]_n} \geqslant x\right) \leqslant \inf_{p>1}\left(\mathbb{E}\left[\exp\left(-\frac{(p-1)x^2}{2}\Big(2ab+b^2[M]_n\Big)\right)\right]\right)^{1/p}, \tag{3.81}$$

$$\mathbb{P}\left(\frac{M_n}{a+b[M]_n} \geqslant x, [M]_n \geqslant y\right) \leqslant \exp\left(-\frac{x^2}{2}\Big(2ab+b^2y\Big)\right). \tag{3.82}$$

Moreover, we also have

$$\mathbb{P}\left(\frac{M_n}{a+b\langle M\rangle_n} \geqslant x, [M]_n \leqslant y\langle M\rangle_n\right) \tag{3.83}$$

$$\leqslant \inf_{p>1}\left(\mathbb{E}\left[\exp\left(-\frac{(p-1)x^2}{2y}\Big(2ab+b^2\langle M\rangle_n\Big)\right)\right]\right)^{1/p}.$$

The proofs of those exponential inequalities rely on the following keystone Lemma which can be compared with Lemma 3.33.

Lemma 3.40. *For a random variable* X *and for any real* t, *denote*

$$L(t) = \mathbb{E}\left[\exp\left(tX - \frac{t^2}{2}X^2\right)\right]. \tag{3.84}$$

1) If X *is heavy on left, then for any positive* t, $L(t) \leqslant 1$.
2) If X *is heavy on right, then for any negative* t, $L(t) \leqslant 1$.
3) If X *is symmetric, then for any real* t, $L(t) \leqslant 1$.

Remark 3.41. A recent work of Delyon [12] reveals that it is possible to get rid of the assumption that $\mathbb{E}[X] = 0$ in Lemma 3.40, which means that Lemma 3.40 holds true without the assumption that $\mathbb{E}[X] = 0$. However, in the martingale framework, it is quite natural to assume that $\mathbb{E}[X] = 0$.

Proof. Let f be the function defined, for all real x, by

$$f(x) = \exp\left(x - \frac{x^2}{2}\right).$$

On the one hand, it follows from the well-known inequality $\cosh(x) \leqslant \exp(x^2/2)$ together with the identity $\cosh(x) = \exp(x)/(1 + \tanh(x))$ that

$$\frac{\exp(x)}{1 + \tanh(x)} \leqslant \exp\left(\frac{x^2}{2}\right)$$

which is equivalent to say that for any real x,

$$f(x) \leqslant 1 + \tanh(x). \tag{3.85}$$

On the other hand, via a simple integration by parts, we have for any real x,

$$\tanh(x) = -\int_0^{+\infty} \tanh''(a) T_a(x)\, da$$

which implies that, for any positive t,

$$\tanh(tx) = -t^2 \int_0^{+\infty} \tanh''(at) T_a(x)\, da.$$

By taking the expectation on both sides of the above identity, we obtain that

$$\mathbb{E}[\tanh(tX)] = -t^2 \int_0^{+\infty} \tanh''(at) \mathbb{E}[T_a(X)]\, da. \tag{3.86}$$

Consequently, if X is heavy on left, we deduce from (3.85) and (3.86) that, for any positive t, $\mathbb{E}[\tanh(tX)] \leqslant 0$, which leads to part 1) of Lemma 3.40,

$$L(t) = \mathbb{E}[f(tX)] \leqslant 1 + \mathbb{E}[\tanh(tX)] \leqslant 1.$$

Furthermore, if X is heavy on right, $-X$ is heavy on left. Hence, we immediately infer from part 1) of Lemma 3.40 that $L(t) \leqslant 1$ for any negative t. Finally, part 3) of Lemma 3.40 follows from the conjunction of parts 1) and 2). $\qquad\square$

Proof of Theorems 3.38 and 3.39. For any positive t and for all $n \geqslant 0$, denote

$$W_n(t) = \exp\left(t M_n - \frac{t^2}{2}[M]_n\right)$$

with $W_0(t) = 1$. We claim that if (M_n) is heavy on left, then for any positive t, $(W_n(t))$ is a supermartingale with $\mathbb{E}[W_n(t)] \leqslant 1$. As a matter of fact, for any positive t and for all $n \geqslant 1$,

$$W_n(t) = W_{n-1}(t) \exp\Big(t\Delta M_n - \frac{t^2}{2}\Delta[M]_n\Big)$$

where $\Delta[M]_n = \Delta M_n^2$. We infer from Lemma 3.40 part 1) that for any positive t and for all $n \geqslant 1$,

$$\mathbb{E}\left[\exp\Big(t\Delta M_n - \frac{t^2}{2}\Delta M_n^2\Big)\Big|\mathscr{F}_{n-1}\right] \leqslant 1.$$

Consequently, for any positive t, $(W_n(t))$ is a positive supermartingale satisfying, for all $n \geqslant 1$, $\mathbb{E}[W_n(t)] \leqslant \mathbb{E}[W_{n-1}(t)]$ leading to $\mathbb{E}[W_n(t)] \leqslant 1$. We are now in position to prove Theorems 3.38 and 3.39. The proofs of (3.80) and (3.81) follow exactly the same lines as those of (3.65) and (3.66), respectively. We shall proceed to the proof of Theorem 3.39. For any positive x and y, denote

$$A_n = \Big\{M_n \geqslant x[M]_n, [M]_n \geqslant y\Big\}.$$

We clearly have

$$\mathbb{P}(A_n) \leqslant \mathbb{E}\left[\exp\Big(xM_n - x^2[M]_n\Big)\mathrm{I}_{A_n}\right],$$
$$\leqslant \exp\Big(-\frac{x^2y}{2}\Big)\mathbb{E}\left[W_n(x)\right],$$
$$\leqslant \exp\Big(-\frac{x^2y}{2}\Big).$$

Moreover, for any positive x and y, let

$$B_n = \Big\{M_n \geqslant x(a+b\langle M\rangle_n), [M]_n \leqslant y\langle M\rangle_n\Big\}.$$

By Hölder's inequality, we have for any positive t and for all $q > 1$,

$$\mathbb{P}(B_n) \leqslant \mathbb{E}\left[\exp\Big(\frac{t}{q}M_n - \frac{tax}{q} - \frac{tbx}{q}\langle M\rangle_n\Big)\mathrm{I}_{B_n}\right],$$
$$\leqslant \exp\Big(-\frac{tax}{q}\Big)\mathbb{E}\left[\exp\Big(\frac{t}{q}M_n - \frac{t^2}{2q}[M]_n\Big)\exp\Big(\frac{t}{2q}(ty - 2bx)\langle M\rangle_n\Big)\mathrm{I}_{B_n}\right],$$
$$\leqslant \exp\Big(-\frac{tax}{q}\Big)\mathbb{E}\left[(W_n(t))^{1/q}\exp\Big(\frac{t}{2q}(ty - 2bx)\langle M\rangle_n\Big)\right],$$
$$\leqslant \exp\Big(-\frac{tax}{q}\Big)\left(\mathbb{E}\left[\exp\Big(\frac{tp}{2q}(ty - 2bx)\langle M\rangle_n\Big)\right]\right)^{1/p} \tag{3.87}$$

where p and q are Hölder conjugate exponents. Consequently, as $p/q = p - 1$, we can deduce from (3.87) with the optimal value $t = bx/y$ that

$$\mathbb{P}(B_n) \leq \inf_{p>1} \left(\mathbb{E}\left[\exp\left(-\frac{(p-1)x^2}{2y} \left(2ab + b^2 \langle M \rangle_n \right) \right) \right] \right)^{1/p}$$

which achieves the proof of Theorem 3.39. $\square$

3.8 Complements and Exercises

Exercise 1 (Linear regression processes, [3]). Consider the stochastic linear regression process given, for all $n \geq 1$, by

$$X_n = \theta \phi_{n-1} + \varepsilon_n$$

where X_n, ϕ_n, and ε_n are the observation, the regression variable, and the driven noise, respectively. We assume that (ϕ_n) and (ε_n) are two independent sequences of independent and identically distributed random variables. We also suppose that (ε_n) is a bounded sequence with zero mean and finite variance $\sigma^2 > 0$. More precisely, we assume that, for all $n \geq 1$, $|\varepsilon_n| \leq b$ a.s. where b is a positive real number. We estimate the unknown parameter θ by the least-squares estimator $\hat{\theta}_n$ given, for all $n \geq 1$, by

$$\hat{\theta}_n = \frac{\sum_{k=1}^n \phi_{k-1} X_k}{\sum_{k=1}^n \phi_{k-1}^2}$$

1) Prove that, for all $n \geq 1$,

$$\hat{\theta}_n - \theta = \sigma^2 \frac{M_n}{\langle M \rangle_n}$$

where

$$M_n = \sum_{k=1}^n \phi_{k-1} \varepsilon_k \quad \text{and} \quad \langle M \rangle_n = \sigma^2 \sum_{k=1}^n \phi_{k-1}^2.$$

2) Deduce from inequality (3.74) that, for any positive x,

$$\mathbb{P}(\hat{\theta}_n - \theta \geq x) \leq \inf_{p>1} \left(\mathbb{E}\left[\exp\left(-\frac{3(p-1)x^2}{2\sigma^2(2\sigma^2 + b^2)} \langle M \rangle_n \right) \right] \right)^{1/p}.$$

3) In the special case where (ϕ_n) shares the same $\mathcal{N}(0,1)$ distribution, prove from the above inequality with $p = 2$ that, for any positive x,

$$\mathbb{P}(\hat{\theta}_n - \theta \geq x) \leq \exp\left(-\frac{n}{4} \log\left(1 + \frac{3x^2}{\sigma^2(2\sigma^2 + b^2)} \right) \right).$$

Exercise 2 (Autoregressive processes). Consider the first-order autoregressive process given, for all $n \geqslant 1$, by

$$X_n = \theta X_{n-1} + \varepsilon_n$$

where the parameter $|\theta| < 1$, the initial state $X_0 = 0$ and the driven noise (ε_n) is a sequence of independent and identically distributed random variables sharing the same Rademacher $\mathscr{R}(1/2)$ distribution, which means that, for all $n \geqslant 1$, ε_n is equal to $+1$ or to -1 with the same probability $1/2$. We estimate the unknown parameter θ by the least-squares estimator given, for all $n \geqslant 1$, by

$$\hat{\theta}_n = \frac{\sum_{k=1}^n X_{k-1} X_k}{\sum_{k=1}^n X_{k-1}^2}.$$

1) Prove that, for all $n \geqslant 1$,

$$\hat{\theta}_n - \theta = \frac{M_n}{\langle M \rangle_n}$$

where (M_n) is the square integrable martingale given by

$$M_n = \sum_{k=1}^n X_{k-1} \varepsilon_k \quad \text{and} \quad [M]_n = \langle M \rangle_n = \sum_{k=1}^n X_{k-1}^2.$$

2) Show that, for any real number t and for all $n \geqslant 1$,

$$\mathbb{E}[\exp(t X_n^2)|\mathscr{F}_{n-1}] = \exp(t + t\theta^2 X_{n-1}^2)\cosh(2t\theta X_{n-1}).$$

3) Deduce from De la Peña's inequality (3.50) that, for any x in $]0,1[$,

$$\mathbb{P}(\hat{\theta}_n - \theta \geqslant x) \leqslant \exp\left(-\frac{(n-1)x^2}{4}\right).$$

Exercise 3 (Galton-Watson processes). Consider the Galton-Watson process starting from $X_0 = 1$ and given, for all $n \geqslant 1$, by

$$X_n = \sum_{k=1}^{X_{n-1}} Y_{n,k}$$

where $(Y_{n,k})$ is a sequence of independent and identically distributed, nonnegative integer-valued random variables. The distribution of $(Y_{n,k})$, with finite mean m and variance σ^2, is commonly called the offspring or reproduction distribution. We assume that $m > 1$ and we estimate the offspring mean m by the Lotka-Nagaev estimator given, for all $n \geqslant 1$, by

$$\hat{m}_n = \frac{X_n}{X_{n-1}}.$$

Without loss of generality, we suppose that the set of extinction of the process (X_n) is negligeable. Hence, the Lokta-Nagaev estimator $\hat{m}_n$ is always well defined.

1) Prove that the Galton-Watson process can be rewritten in the autoregressive form

$$X_n = mX_{n-1} + \varepsilon_n$$

where $\mathbb{E}[\varepsilon_n | \mathscr{F}_{n-1}] = 0$ and $\mathbb{E}[\varepsilon_n^2 | \mathscr{F}_{n-1}] = \sigma^2 X_{n-1}$.

2) For all $n \geqslant 1$, let

$$M_n = \frac{X_n}{m^n}.$$

Show that (M_n) is a martingale bounded in $\mathbb{L}^2$ which converges a.s. and in $\mathbb{L}^2$ to a positive random variable M, and compute the mean and the variance of M.

3) Prove that

$$\lim_{n \to \infty} \hat{m}_n = m \qquad \text{a.s.}$$

4) Denote by L the log-Laplace transform of the centered offspring distribution given, for any real number t, by $L(t) = \log \mathbb{E}[\exp(t(Y_{n,k} - m))]$. Show that, for any real number t,

$$\mathbb{E}[\exp(t\varepsilon_n) | \mathscr{F}_{n-1}] = \exp(L(t)X_{n-1}).$$

5) Assume that L is finite in a right neighborhood of the origin and let I be the Legendre-Fenchel transform of L. Establish that, for any positive x,

$$\mathbb{P}(\hat{m}_n - m \geqslant x) \leqslant \mathbb{E}[\exp(-I(x)X_{n-1})].$$

6) In the special case where the offspring distribution is the geometric $\mathscr{G}(p)$ distribution with $0 < p < 1$, prove that, for all $n \geqslant 1$ and for any positive x,

$$\mathbb{P}(\hat{m}_n - m \geqslant x) \leqslant \frac{p^{(n-1)}}{\exp(I(x)) - 1}.$$

Exercise 4 (Random walk, [5]). Consider the random walk on the integer number line $\mathbb{Z}$, which starts from the origin at time 0 and, at each step, moves to the right $+1$ or to the left -1 with equal probability $1/2$. Denote by $(A(n))$ the sequence of random subsets of $\mathbb{Z}$, recursively defined as follows: $A(0) = \{0\}$ and, for all $n \geqslant 0$,

$$A(n+1) = \begin{cases} A(n) \cup \{L_n - 1\} \\ A(n) \cup \{R_n + 1\} \end{cases}$$

if the random walk leaves $A(n)$ by the left side or by the right side, respectively, where L_n and R_n stand for the minimum and the maximum of $A(n)$. To be more precise,

$$A(n) = \{L_n, L_n + 1, \ldots, R_n - 1, R_n\}.$$

The cardinal of $A(n)$ is always $n + 1$ and its length $R_n - L_n = n$. The random set $A(n)$ is characterized by $X_n = L_n + R_n$ as

$$L_n = \frac{X_n - n}{2} \qquad \text{and} \qquad R_n = \frac{X_n + n}{2}.$$

We are interested in the asymptotic behavior of the process (X_n).

1) Use a stopping time argument for gambler's ruin to prove that, for all $n \geqslant 0$,

$$\mathbb{P}(X_{n+1} = X_n - 1 | X_n) = \frac{n + 2 + X_n}{2(n+2)},$$

$$\mathbb{P}(X_{n+1} = X_n + 1 | X_n) = \frac{n + 2 - X_n}{2(n+2)}. \tag{3.88}$$

2) Deduce from the 1) that, for all $n \geqslant 0$,

$$\mathbb{E}[X_{n+1} | X_n] = \frac{(n+1)X_n}{n+2} \qquad \text{and} \qquad \mathbb{E}[X_{n+1}^2 | X_n] = 1 + \frac{nX_n^2}{n+2}.$$

3) Let (M_n) be the sequence defined by $M_n = (n+1)X_n$. Show that (M_n) is a square integrable martingale satisfying

$$\langle M \rangle_n = \sum_{k=1}^{n} (k+1)^2 - \sum_{k=1}^{n} X_{k-1}^2.$$

4) Infer from the strong law of large numbers for martingales given in Theorem 1.18 that $M_n = o(n^2)$ a.s., which means that

$$\lim_{n \to \infty} \frac{X_n}{n} = 0 \qquad \text{a.s.}$$

5) Deduce from Toeplitz's lemma together with 3) and 4) that

$$\lim_{n \to \infty} \frac{\langle M \rangle_n}{n^3} = \frac{1}{3} \qquad \text{a.s.}$$

6) Prove from the central limit theorem for martingales given in Theorem 1.20 that

$$\frac{X_n}{\sqrt{n}} \xrightarrow{\mathcal{L}} \mathcal{N}\left(0, \frac{1}{3}\right).$$

7) Show that, for all $n \geqslant 1$, the increments $|\Delta M_n| \leqslant 2n$ a.s.

8) Conclude from Hoeffding's inequality (3.3) that, for any positive x,

$$\mathbb{P}\left(\frac{|X_n|}{n} \geqslant x\right) \leqslant 2\exp\left(-\frac{3}{8}nx^2\right).$$

Exercise 5. Let $X_1, \ldots, X_n$ be a finite sequence of independent random variables such that, for all $1 \leqslant k \leqslant n$, X_k has the Bernoulli $\mathcal{B}(p_k)$ distribution where $p_k \leqslant 1/2$. Denote

$$M_n = \sum_{k=1}^{n} \frac{p_k - X_k}{p_k}.$$

Prove that, for any positive x,

$$P(M_n \geqslant x) \leqslant \exp\left(-\frac{x^2}{2\mathrm{Var}(M_n)}\right).$$

Exercise 6 (Martingale triangular array). Let $(X_{n,k})_{1\leqslant k\leqslant k_n}$ be a triangular array of integrable random variables defined on the same probability space $(\Omega,\mathscr{A},\mathbb{P})$ and associated with the filtration $(\mathscr{F}_{n,k})_{0\leqslant k\leqslant k_n}$. Assume that, for every integer $n \geqslant 1$ and for all $1 \leqslant k \leqslant k_n$, $\mathbb{E}[X_{n,k}|\mathscr{F}_{n,k-1}] = 0$ and denote

$$M_{n,k} = \sum_{\ell=1}^{k} X_{n,\ell}.$$

The sequence $(M_{n,k})_{1\leqslant k\leqslant k_n}$ is called a martingale triangular array. We assume that, for some constant $a > 0$, the Lyapunov condition is satisfied

$$n^a \sum_{k=1}^{k_n} \mathbb{E}[X_{n,k}^2|\mathscr{F}_{n,k-1}] \overset{\mathcal{P}}{\longrightarrow} 0.$$

In addition, we also suppose that there exists a positive constant b such that, for every $n \geqslant 1$ and for all $1 \leqslant k \leqslant k_n$, $X_{n,k} \leqslant b$ a.s.

1) Prove from Theorem 3.10 that, for any $0 < x < k_n b$ and for any positive y,

$$\mathbb{P}(M_{n,k_n} \geqslant x, \langle M_n\rangle_{k_n} \leqslant y) \leqslant \exp\left(-\frac{y+bx}{b^2+y/k_n}\log\left(1+\frac{bx}{y}\right) - k_n\log\left(1-\frac{x}{k_n b}\right)\right).$$

2) Prove that, for any $x > b/a$,

$$\sum_{n=1}^{\infty} \mathbb{P}(M_{n,k_n} \geqslant x, \langle M_n\rangle_{k_n} \leqslant n^{-a}) < \infty.$$

3) Deduce from the Borel-Cantelli lemma that

$$\varlimsup_{n} M_{n,k_n} \leqslant \frac{b}{a} \qquad \text{a.s.}$$

Exercise 7. Let $\varepsilon_1,\ldots,\varepsilon_n$ be a finite sequence of independent random variables sharing the same Rademacher $\mathscr{R}(1/2)$ distribution, which means that, for all $n \geqslant 1$, ε_n is equal to $+1$ or to -1 with the same probability $1/2$. Let $(a(i,j))$ be an $n \times n$ array of real numbers and denote

$$M_n = \sum_{i=2}^{n}\sum_{j=1}^{i-1} a(i,j)\varepsilon_i\varepsilon_j.$$

1) Use Hoeffding's inequality (3.3) to prove that, for any positive x,

$$\mathbb{P}(M_n \geqslant x) \leqslant \exp\left(-\frac{x^2}{2\mathscr{A}_n}\right)$$

where

$$\mathscr{A}_n = \sum_{i=2}^{n}\left(\sum_{j=1}^{i-1}|a(i,j)|\right)^2.$$

2) Let A be the lower triangular matrix defined by $A_{ij} = a(i,j)$ if $1 \leqslant j < i \leqslant n$ and $A_{ij} = 0$ otherwise. Prove that, for any positive x,

$$\mathbb{P}(M_n \geqslant x) \leqslant \exp\left(-\frac{x^2}{2n\|A\|_{2,2}^2}\right)$$

where

$$\|A\|_{2,2} = \sup\left\{\|Ax\|_2,\ x \in \mathbb{R}^n,\ \|x\|_2 = 1\right\}.$$

Hint: Apply inequality (3.25). Compare the two above inequalities.

Exercise 8 (Martingales with conditionally symmetric differences, [16]). Let (M_n) be a martingale with bounded increments, such that $M_0 = 0$. Assume that for all $n \geqslant 1$, $|\Delta M_n| \leqslant 1$ a.s. and that the distribution of ΔM_n given $\mathscr{F}_{n-1}$ is symmetric.

1) Prove that, for all $n \geqslant 1$ and for any positive t,

$$\log \mathbb{E}[\exp(t\Delta M_n) \mid \mathscr{F}_{n-1}] \leqslant \log(1 + \Delta\langle M\rangle_n(\cosh(t) - 1)),$$
$$\leqslant \Delta\langle M\rangle_n(\cosh(t) - 1).$$

2) Let φ be the function defined, for any positive x, by

$$\varphi(x) = x\log\left(\sqrt{1+x^2}+x\right) + 1 - \sqrt{1+x^2}.$$

Prove that, for any positive x and y,

$$\mathbb{P}(M_n \geqslant x, \langle M\rangle_n \leqslant y) \leqslant \exp\left(-y\varphi\left(\frac{x}{y}\right)\right).$$

Exercise 9 (McDiarmid's inequality, [13]). Let $(E_1, d_1), \ldots, (E_n, d_n)$ be a finite sequence of countable metric spaces with respective finite diameters $c_1, \ldots, c_n$. Denote E^n the product space $E^n = E_1 \times \cdots \times E_n$. Let f be a separately 1-Lipschitz function from E^n into $\mathbb{R}$, as defined in Subsection 2.8. Let $X_1, \ldots, X_n$ be a finite sequence of independent random variables such that, for all $1 \leqslant k \leqslant n$, X_k takes its values in E_k, and denote $Z = f(X_1, \ldots, X_n)$. For all $1 \leqslant k \leqslant n$, let

$$M_k = \sum_{i=1}^{k} \mathbb{E}[Z|\mathscr{F}_i] - \mathbb{E}[Z|\mathscr{F}_{i-1}]$$

where $\mathscr{F}_0 = \{\emptyset, \Omega\}$ and $\mathscr{F}_i = \sigma(X_1, \ldots, X_i)$. One can observe from a standard telescopic argument that $M_n = Z - \mathbb{E}[Z]$.

1) Prove that (M_k) is a martingale with bounded increments.
2) For all $1 \leqslant k \leqslant n$, denote

$$A_k = \inf_{x \in E_k} f(X_1, \ldots, X_{k-1}, x, X_{k+1}, \ldots, X_n),$$

$$B_k = \sup_{x \in E_k} f(X_1, \ldots, X_{k-1}, x_k, X_{k+1}, \ldots, X_n).$$

Prove that, for all $1 \leqslant k \leqslant n$, $B_k - A_k \leqslant c_k$ almost surely.
3) For all $1 \leqslant k \leqslant n$, let $\mathscr{A}_k = \mathbb{E}[A_k | \mathscr{F}_k]$ and $\mathscr{B}_k = \mathbb{E}[B_k | \mathscr{F}_k]$. Show that the random variables $\mathscr{A}_k$ and $\mathscr{B}_k$ are $\mathscr{F}_{k-1}$-measurable, such that $\mathscr{B}_k - \mathscr{A}_k \leqslant c_k$ almost surely.
4) Prove that, for all $1 \leqslant k \leqslant n$, $\mathscr{A}_k \leqslant \Delta M_k \leqslant \mathscr{B}_k$ almost surely.
5) Conclude from Hoeffding's inequality that, for any positive x,

$$\mathbb{P}(Z - \mathbb{E}[Z] \geqslant x) \leqslant \exp\left(-\frac{2x^2}{C_n}\right) \quad \text{where} \quad C_n = \sum_{k=1}^{n} c_k^2.$$

Exercise 10 (Kearns-Saul's inequality for Bernoulli random variables). The aim of this exercise is to obtain an improvement of McDiarmid's inequality in the case of binary random variables. Consider the product space $E^n = \{0,1\}^n$. Let $c_1, \ldots, c_n$ be positive constants and denote by f a function from E^n into $\mathbb{R}$, such that

$$|f(x_1, \ldots, x_n) - f(y_1, \ldots, y_n)| \leqslant \sum_{k=1}^{n} c_k \mathrm{I}_{x_k \neq y_k}$$

for any $x = (x_1, \ldots, x_n)$ and any $y = (y_1, \ldots, y_n)$ in E^n. Let $X_1, \ldots, X_n$ be a finite sequence of independent random variables such that, for all $1 \leqslant k \leqslant n$, X_k has the Bernoulli $\mathscr{B}(p_k)$ distribution where $0 < p_k < 1$. Denote $Z = f(X_1, \ldots, X_n)$ and define $M_1, \ldots, M_n$ and the filtration $\mathscr{F}_0, \ldots, \mathscr{F}_n$ as in the previous exercise. Moreover, for all $1 \leqslant k \leqslant n$, denote by $\mathscr{F}^{(k)}$ the σ-algebra generated by $X_1, \ldots, X_n$ except X_k,

$$\mathscr{F}^{(k)} = \sigma(X_1, \ldots, X_{k-1}, X_{k+1}, \ldots, X_n).$$

1) Using Lemma 2.36, prove that for any positive t and for all $1 \leqslant k \leqslant n$,

$$\log \mathbb{E}[\exp(tZ)|\mathscr{F}^{(k)}] \leqslant t\mathbb{E}[Z|\mathscr{F}^{(k)}] + \frac{c_k^2}{4}\varphi(p_k)t^2$$

where φ is the function defined, for all p in $]0,1[$, by

$$\varphi(p) = \begin{cases} \dfrac{|1-2p|}{|\log((1-p)/p)|} & \text{if } p \neq \tfrac{1}{2}, \\ \dfrac{1}{2} & \text{if } p = \tfrac{1}{2}. \end{cases}$$

2) Deduce that, for any positive t and for all $1 \leqslant k \leqslant n$,

$$\log \mathbb{E}[\exp(t \Delta M_k) | \mathscr{F}_{k-1}] \leqslant \frac{c_k^2}{4} \varphi(p_k) t^2.$$

3) Prove that, for any positive x,

$$\mathbb{P}(Z - \mathbb{E}[Z] \geqslant x) \leqslant \exp\left(-\frac{x^2}{\mathscr{P}_n}\right) \quad \text{where} \quad \mathscr{P}_n = \sum_{k=1}^{n} c_k^2 \varphi(p_k).$$

4) Compare the above inequality with McDiarmid's inequality.
5) Let p be any real number in $]0, 1/2[$ and assume that we are in the particular situation where, for all $1 \leqslant k \leqslant n$, $p_k = p$. Prove that, for any positive x,

$$\mathbb{P}(Z - \mathbb{E}[Z] \geqslant \|c\|_2 x) \leqslant \left(\frac{p}{1-p}\right)^{(1-2p)x^2} \quad \text{where} \quad \|c\|_2^2 = \sum_{k=1}^{n} c_k^2.$$

References

1. Azuma, K.: Weighted sums of certain dependent random variables. Tohoku Math. J. **19**, 357–367 (1967)
2. Barlow, M. T., Jacka, S. D. and Yor, M.: Inequalities for a pair of processes stopped at a random time. Proc. London Math. Soc. **52**, 142–172 (1986)
3. Bercu, B. and Touati, A.: Exponential inequalities for self-normalized martingales with applications. Ann. Appl. Probab. **18**, 1848–1869 (2008)
4. Bernstein, S. Sur quelques modifications de l'inégalité de Tchebycheff. C.R. (Doklady) Acad. Sci. URSS **17**, 279–282 (1937)
5. Chafaï, D. and Malrieu, F.: Recueil de modèles stochastiques. (2015)
6. De la Peña, V. H.: A general class of exponential inequalities for martingales and ratios. Ann. Probab. **27**, 537–564 (1999)
7. De la Peña, V. H., Klass, M. J. and Lai, T. L.: Self-normalized processes: exponential inequalities, moment bounds and iterated logarithm laws. Ann. Probab. **32**, 1902–1933 (2004)
8. De la Peña, V. H., Klass, M. J. and Lai, T. L.: Pseudo-maximization and self-normalized processes. Probab. Surv. **4**, 172–192 (2007)
9. De la Peña, V. H. and Pang, G.: Exponential inequalities for self-normalized processes with applications. Electron. Commun. Probab. **14**, 372–381 (2009)
10. De la Peña, V. H., T. L. and Shao, Q.-M.: Self-normalized processes. Springer-Verlag, Berlin (2009)
11. Delyon, B.: Exponential inequalities for sums of weakly dependent variables. Electron. J. Probab. **14**, 752–779 (2009)
12. Delyon, B.: Exponential inequalities for dependent processes. Preprint hal-01072019 (2015)
13. Devroye, L. and Lugosi, G.: Combinatorial Methods in Density Estimation. Springer-Verlag, New York (2001)
14. Dzhaparidze, K. and van Zanten, J. H.: On Bernstein-type inequalities for martingales. Stochastic Process. Appl. **93**, 109–117 (2001)
15. Fan, X., Grama, I. and Liu, Q.: Hoeffding's inequality for supermartingales. Stochastic Process. Appl. **122**, 3545–3559 (2012)
16. Fan, X., Grama, I. and Liu, Q.: Exponential inequalities for martingales with applications. Electron. J. Probab. **20**, 1–22 (2015)

17. Freedman, D. A.: On tail probabilities for martingales. Ann. Probab. **3**, 100–118 (1975)
18. Hoeffding, W.: Probability inequalities for sums of bounded random variables. J. Amer. Statist. Assoc. **58**, 13–30 (1963)
19. Kearns, M. J. and Saul, L. K.: Large deviation methods for approximate probabilistic inference. Proceedings of the 14th Conference on Uncertaintly in Artifical Intelligence, San-Francisco, 311–319 (1998)
20. Pinelis, I.: Optimum bounds for the distributions of martingales in Banach space. Ann. Probab. **22**, 1679–1706 (1994)
21. Van de Geer, S.: On Hoeffding's inequality for dependent random variables. Empirical process techniques for dependent data. Birkhäuser, Boston, 161–169 (2002)

Chapter 4
Applications in probability and statistics

4.1 Autoregressive process

Our first application is devoted to parameter estimation for first-order autoregressive processes. Autoregressive processes play an important role in many areas of applied mathematics such as probability, statistics, and econometrics. For example, they are widely used in econometrics for forecasting the variability of stock market indices. Many theoretical results were established on parameter estimation for autoregressive processes. However, very few references are available on large deviations or concentration inequalities for autoregressive processes. Consider the first-order autoregressive process given, for all $n \geqslant 1$, by

$$X_n = \theta X_{n-1} + \varepsilon_n \tag{4.1}$$

where X_n and ε_n are the observation and the driven noise, respectively. We assume that (ε_n) is a sequence of independent and identically distributed random variables with standard $\mathcal{N}(0, \sigma^2)$ distribution where $\sigma^2 > 0$. The process is said to be stable if $|\theta| < 1$, unstable if $|\theta| = 1$, and explosive if $|\theta| > 1$. We estimate the unknown parameter θ by the least-squares estimator given, for all $n \geqslant 1$, by

$$\hat{\theta}_n = \frac{\sum_{k=1}^n X_{k-1} X_k}{\sum_{k=1}^n X_{k-1}^2}. \tag{4.2}$$

In the stable case $|\theta| < 1$, it is well known that $\hat{\theta}_n$ converges almost surely to θ. In addition, we also have the asymptotic normality

$$\sqrt{n}(\hat{\theta}_n - \theta) \xrightarrow{\mathcal{L}} \mathcal{N}(0, 1 - \theta^2).$$

© The Authors 2015

B. Bercu et al., *Concentration Inequalities for Sums and Martingales*,
SpringerBriefs in Mathematics, DOI 10.1007/978-3-319-22099-4_4

The large deviation properties of the sequence $(\hat{\theta}_n)$ were established in [3]. More precisely, let

$$a = \frac{\theta - \sqrt{\theta^2 + 8}}{4} \qquad \text{and} \qquad b = \frac{\theta + \sqrt{\theta^2 + 8}}{4}.$$

Assume that $|\theta| < 1$ and that X_0 is independent of (ε_n) with $\mathcal{N}(0, \sigma^2/(1 - \theta^2))$ distribution. Then, the sequence $(\hat{\theta}_n)$ satisfies a large deviation principle with a convex–concave rate function given by

$$I(x) = \begin{cases} \dfrac{1}{2} \log\left(\dfrac{1 + \theta^2 - 2\theta x}{1 - x^2} \right) & \text{if } x \in [a,b], \\ \log | \theta - 2x | & \text{otherwise.} \end{cases}$$

We shall now propose, whatever the value of θ is, a very simple exponential inequality for the estimator $\hat{\theta}_n$. We refer the reader to [4] for more details on this result.

Theorem 4.1. *For all $n \geqslant 1$ and for any positive x,*

$$\mathbb{P}(|\hat{\theta}_n - \theta| \geqslant x) \leqslant 2\exp\left(-\frac{nx^2}{2(1 + y_x)}\right) \tag{4.3}$$

where y_x is the unique positive solution of the equation $h(y_x) = x^2$ and h is the function given, for any positive x, by $h(x) = (1 + x)\log(1 + x) - x$.

Remark 4.2. Inequality (4.3) can be very simple for x small enough. As a matter of fact, one can easily see that for any $0 < x < 1$, $h(x) > x^2/4$. Consequently, it immediately follows from (4.3) that, for any $0 < x < 1/2$,

$$\mathbb{P}(|\hat{\theta}_n - \theta| \geqslant x) \leqslant 2\exp\left(-\frac{nx^2}{2(1 + 2x)}\right).$$

We refer to inequality (A.15) of Del Moral and Rio [9] for a more precise upper bound on y_x.

Proof. It immediately follows from (4.1) together with (4.2) that for all $n \geqslant 1$,

$$\hat{\theta}_n - \theta = \sigma^2 \frac{M_n}{\langle M \rangle_n} \tag{4.4}$$

where

$$M_n = \sum_{k=1}^{n} X_{k-1} \varepsilon_k \qquad \text{and} \qquad \langle M \rangle_n = \sigma^2 \sum_{k=1}^{n} X_{k-1}^2.$$

The driven noise (ε_n) is a sequence of independent and identically distributed random variables with $\mathcal{N}(0, \sigma^2)$ distribution. Consequently, for all $n \geqslant 1$, the distribution of the increments $\Delta M_n = X_{n-1}\varepsilon_n$ given $\mathscr{F}_{n-1}$ is $\mathcal{N}(0, \sigma^2 X_{n-1}^2)$ which means that (M_n) is a Gaussian martingale. Therefore, we infer from inequality (3.66) that for all $n \geqslant 1$ and for any positive x,

$$\mathbb{P}(|\hat{\theta}_n - \theta| \geqslant x) = \mathbb{P}\left(|M_n| \geqslant \frac{x}{\sigma^2}\langle M\rangle_n\right) = 2\,\mathbb{P}\left(M_n \geqslant \frac{x}{\sigma^2}\langle M\rangle_n\right),$$

$$\leqslant 2\inf_{p>1}\left(\mathbb{E}\left[\exp\left(-(p-1)\frac{x^2}{2\sigma^4}\langle M\rangle_n\right)\right]\right)^{1/p}. \tag{4.5}$$

Similar result may be found in [15–17]. It only remains to find a suitable upper-bound for the right-hand side in (4.5). For any real t such that $1 - 2\sigma^2 t > 0$, let

$$\alpha = \frac{1}{\sqrt{1 - 2\sigma^2 t}}.$$

We deduce from (4.1) that, for all $n \geqslant 1$,

$$\mathbb{E}[\exp(tX_n^2)|\mathscr{F}_{n-1}] = \exp(t\theta^2 X_{n-1}^2)\mathbb{E}[\exp(2\theta t X_{n-1}\varepsilon_n + t\varepsilon_n^2)|\mathscr{F}_{n-1}],$$

$$= \frac{\exp(t\theta^2 X_{n-1}^2)}{\sigma\sqrt{2\pi}}\int_{\mathbb{R}}\exp\left(-\frac{x^2}{2\alpha^2\sigma^2}\right)\exp(2\theta t X_{n-1}x)dx.$$

Hence, if $\beta = 2t\alpha\sigma\theta X_{n-1}$, we find via the change of variables $y = x/\alpha\sigma$ that

$$\mathbb{E}[\exp(tX_n^2)|\mathscr{F}_{n-1}] = \frac{\alpha\exp(t\theta^2 X_{n-1}^2)}{\sqrt{2\pi}}\int_{\mathbb{R}}\exp\left(-\frac{y^2}{2} + \beta y\right)dy,$$

$$= \alpha\exp\left(t\theta^2 X_{n-1}^2 + \frac{\beta^2}{2}\right) = \alpha\exp(t\alpha^2\theta^2 X_{n-1}^2),$$

which implies that, for any negative t and for all $n \geqslant 1$,

$$\mathbb{E}[\exp(tX_n^2)|\mathscr{F}_{n-1}] \leqslant \alpha. \tag{4.6}$$

Furthermore, we also have $\mathbb{E}[\exp(tX_0^2)] \leqslant \alpha$ as X_0 is $\mathscr{N}(0, \sigma^2/(1-\theta^2))$ distributed. Consequently, via the tower property of the conditional expectation, we find from (4.6) that for any negative t and for all $n \geqslant 0$,

$$\mathbb{E}[\exp(t\langle M\rangle_n)] \leqslant (1 - 2\sigma^4 t)^{-n/2}. \tag{4.7}$$

Then, we obtain from (4.5) and (4.7), with the optimal value $t = -(p-1)x^2/2\sigma^4$, that for any positive x and for all $n \geqslant 1$,

$$\mathbb{P}(|\hat{\theta}_n - \theta| \geqslant x) \leqslant 2\inf_{p>1}\exp\left(-\frac{n}{2p}\log(1 + (p-1)x^2)\right).$$

Hence, the change of variables $y = (p-1)x^2$ leads, for all $n \geqslant 1$, to

$$\mathbb{P}(|\hat{\theta}_n - \theta| \geqslant x) \leqslant 2\inf_{y>0}\exp\left(-\frac{nx^2}{2}\ell(y)\right)$$

where the function ℓ is given, for any positive y, by

$$\ell(y) = \frac{\log(1+y)}{x^2+y}.$$

We clearly have

$$\ell'(y) = \frac{x^2 - h(y)}{(1+y)(x^2+y)^2}$$

where $h(y) = (1+y)\log(1+y) - y$. One can observe that the function h is the Cramer transform of the centered Poisson distribution with parameter $\lambda = 1$. Let y_x be the unique positive solution of the equation $h(y_x) = x^2$. The value y_x maximizes the function ℓ and this natural choice clearly leads to (4.3), which completes the proof of Theorem 4.1. $\square$

4.2 Random permutations

In 1951, Wassily Hoeffding [13] raised the following question, motivated by a problem of nonparametric testing of independence. Let $(a_n(i,j))$ be an $n \times n$ array of real numbers. Let π_n be chosen uniformly at random from the set of all permutations of $\{1,\dots,n\}$, and denote

$$S_n = \sum_{i=1}^{n} a_n(i, \pi_n(i)).$$

Under which reasonable conditions on the array $(a_n(i,j))$, does the sequence (S_n) suitably standardized, converge in distribution to the standard normal distribution? It is straightforward to see that

$$\mathbb{E}[S_n] = \frac{1}{n}\sum_{i=1}^{n}\sum_{j=1}^{n} a_n(i,j) \quad \text{and} \quad \operatorname{Var}(S_n) = \frac{1}{n-1}\sum_{i=1}^{n}\sum_{j=1}^{n} d_n^2(i,j)$$

where $d_n(i,j) = a_n(i,j) - a_n(i,*) - a_n(*,j) + a_n(*,*)$ with

$$a_n(i,*) = \frac{1}{n}\sum_{\ell=1}^{n} a_n(i,\ell), \quad a_n(*,j) = \frac{1}{n}\sum_{k=1}^{n} a_n(k,j),$$

$$a_n(*,*) = \frac{1}{n^2}\sum_{k=1}^{n}\sum_{\ell=1}^{n} a_n(k,\ell).$$

Moreover, it was shown in [13] that, as soon as

$$\max_{1 \leqslant i,j \leqslant n} d_n^2(i,j) = o\big(\operatorname{Var}(S_n)\big),$$

then

$$\frac{S_n - \mathbb{E}[S_n]}{\sqrt{\mathrm{Var}(S_n)}} \xrightarrow{\mathcal{L}} \mathcal{N}(0,1).$$

Our objective is now to provide concentration inequality for $S_n - \mathbb{E}[S_n]$.

Theorem 4.3. *Let $(a_n(i,j))$ be an $n \times n$ array of real numbers from $[-m_a, m_a]$. Let*

$$S_n = \sum_{i=1}^{n} a_n(i, \pi_n(i))$$

where π_n is drawn from the uniform distribution over the set of all permutations of $\{1, \ldots, n\}$. Then, for any positive x,

$$\mathbb{P}(|S_n - \mathbb{E}[S_n]| \geqslant x) \leqslant 4 \exp\left(-\frac{x^2}{16(\theta v_n + x m_a/3)}\right) \tag{4.8}$$

where $\theta = \dfrac{5}{2}\log(3) - \dfrac{2}{3}$ and

$$v_n = \frac{1}{n} \sum_{i-1}^{n} \sum_{j-1}^{n} a_n^2(i,j). \tag{4.9}$$

Remark 4.4. It is easy to check that the sum

$$\Sigma_n = \sum_{i=1}^{n} d_n(i, \pi_n(i))$$

satisfies $\Sigma_n = S_n - \mathbb{E}[S_n]$. Moreover, an application of our result to Σ_n leads to

$$\mathbb{P}(|S_n - E[S_n]| \geqslant x) \leqslant 4 \exp\left(-\frac{x^2}{16(\theta \mathrm{Var}(S_n) + 4x m_a/3)}\right).$$

Remark 4.5. It was proven by Chatterjee [7], in the special case $0 \leqslant a_n(i,j) \leqslant m_a$, that for any positive x,

$$\mathbb{P}(|S_n - E[S_n]| \geqslant x) \leqslant 2 \exp\left(-\frac{x^2}{4 m_a \mathbb{E}[S_n] + 2x m_a}\right).$$

One can observe that this upper bound has better constants. However, the variance term v_n has been replaced with a drastic bound $m_a \mathbb{E}[S_n]$.

In order to prove Theorem 4.3, we need the following result which contains an unfortunate logarithmic factor. We shall get rid later of this factor by a splitting argument. The unnecessary indices will henceforth be omitted.

Lemma 4.6. *Let $(a(i,j))$ be an $p \times n$ array of real numbers from $[-m_a, m_a]$ with $1 \leqslant p \leqslant n$ and denote*

$$S_p = \sum_{i=1}^{p} a(i, \pi(i))$$

where π is drawn from the uniform distribution over the set of all one-to-one mappings from $\{1,\ldots p\}$ to $\{1,\ldots n\}$. Then, for any positive x,

$$\mathbb{P}(|S_p - \mathbb{E}[S_p]| \geqslant x) \leqslant 2\exp\left(-\frac{x^2}{4(\theta v + 2xm_a/3)}\right) \tag{4.10}$$

where

$$v = \frac{1}{n}\sum_{i=1}^{p}\sum_{j=1}^{n} a^2(i,j),$$

$$\alpha = \frac{p}{n} \qquad and \qquad \theta = -\alpha - \frac{1+\alpha}{\alpha}\log(1-\alpha).$$

Proof. The sequence π is recursively defined through the Knuth algorithm. First, $\pi(1)$ is uniformly drawn in $\{1,\ldots n\}$. Next, for $i = 2$ to $i = p$, $\pi(i)$ is uniformly drawn in I_{i-1} where

$$I_i = \{1,\ldots n\}\backslash\{\pi(1),\ldots \pi(i)\}.$$

It is well known that if we pursue down to $i = n$, we obtain a uniformly generated permutation of $\{1,\ldots n\}$. By stopping at $i = p$, the algorithm has thus generated uniformly a one-to-one mapping from $\{1,\ldots,p\}$ to $\{1,\ldots,n\}$. Let $\mathscr{F}_0$ be the trivial σ-algebra and, for $1 \leqslant i \leqslant p$, define

$$\mathscr{F}_i = \sigma\{\pi(1),\ldots \pi(i)\}.$$

Let $M_1,\ldots,M_p$ be the finite sequence defined, for all $1 \leqslant i \leqslant p$, by

$$M_i = \mathbb{E}[S_p|\mathscr{F}_i] - \mathbb{E}[S_p]$$

and $M_0 = \mathbb{E}[S_p|\mathscr{F}_0] - \mathbb{E}[S_p] = 0$. It follows from the tower property of the conditional expectation that, for all $1 \leqslant i \leqslant p$,

$$\mathbb{E}[\Delta M_i|\mathscr{F}_{i-1}] = \mathbb{E}[M_i - M_{i-1}|\mathscr{F}_{i-1}] = \mathbb{E}[S_p|\mathscr{F}_{i-1}] - \mathbb{E}[S_p|\mathscr{F}_{i-1}] = 0.$$

Consequently, (M_i) is a martingale such that $M_p = S_p - \mathbb{E}[S_p]$. Moreover, we claim that, for all $1 \leqslant i \leqslant p$,

$$|\Delta M_i| \leqslant b \qquad \text{a.s.} \tag{4.11}$$

where $b = 4m_a$. Consequently, we deduce from (3.31) that, for any positive x,

$$\mathbb{P}(|S_p - \mathbb{E}[S_p]| \geqslant x) \leqslant 2\exp\left(-\frac{x^2}{2(y + bx/3)}\right) \tag{4.12}$$

where $y = \|\langle M \rangle_p\|_\infty$. It remains to proceed to the proof of (4.11) and to find a suitable upper bound for the increasing process $\langle M \rangle_p$. For any $1 \leqslant i \leqslant p$, denote

$$A_i = \sum_{j=1}^{n} a^2(i,j) \qquad \text{and} \qquad \mathscr{B} = \sum_{i=1}^{p} A_i.$$

We can assume, without loss of generality, that the indices are ordered in such a way that the sequence (A_i) is nonincreasing. For all $1 \leqslant i \leqslant p-1$, we have

$$M_i = \sum_{j=1}^{i} a(j, \pi(j)) + \sum_{j=i+1}^{p} \mathbb{E}[a(j, \pi(j))|\mathscr{F}_i] - \mathbb{E}[S_p],$$

$$= \sum_{j=1}^{i} a(j, \pi(j)) + \sum_{j=i+1}^{p} m_i(a(j, *)) - \mathbb{E}[S_p]$$

where

$$m_i(a(j, *)) = \frac{1}{n-i} \sum_{k \in I_i} a(j,k).$$

Consequently, for all $1 \leqslant i \leqslant p-1$,

$$\Delta M_i = a(i, \pi(i)) - m_{i-1}(a(i, *)) + \sum_{j=i+1}^{p} m_i(a(j, *)) - m_{i-1}(a(j, *)).$$

However, $I_i = I_{i-1} \setminus \{\pi(i)\}$ which means that $I_{i-1} = I_i \cup \{\pi(i)\}$. Hence,

$$(n-i)m_i(a(j, *)) = \sum_{k \in I_i} a(j,k) = -a(j, \pi(i)) + \sum_{k \in I_{i-1}} a(j,k),$$

$$= -a(j, \pi(i)) + (n-i+1)m_{i-1}(a(j, *)),$$

which ensures that

$$m_i(a(j, *)) - m_{i-1}(a(j, *)) = \frac{1}{n-i}\left(-a(j, \pi(i)) + m_{i-1}(a(j, *))\right).$$

Therefore, we find that for all $1 \leqslant i \leqslant p-1$,

$$\Delta M_i = U_i - \mathbb{E}[U_i|\mathscr{F}_{i-1}] \tag{4.13}$$

where

$$U_i = a(i, \pi(i)) - \frac{1}{n-i} \sum_{j=i+1}^{p} a(j, \pi(i)).$$

In addition, we already saw that $M_p = S_p - \mathbb{E}[S_p]$, so $\Delta M_p = U_p - \mathbb{E}[U_p|\mathscr{F}_{p-1}]$ where $U_p = a(p, \pi(p))$. We are now in position to prove (4.11). For any $1 \leqslant i \leqslant p$ and $1 \leqslant j \leqslant p$, $|a(j, \pi(i))| \leqslant m_a$ a.s. Consequently,

$$|U_i| \leqslant m_a\left(1 + \frac{p-i}{n-i}\right) \leqslant m_a\left(1 + \frac{p}{n}\right) \leqslant 2m_a \qquad \text{a.s.} \qquad (4.14)$$

Then, (4.11) clearly follows from (4.13) and (4.14). Furthermore, we have for all $1 \leqslant i \leqslant p$, $\mathbb{E}[\Delta M_i^2|\mathscr{F}_{i-1}] \leqslant \mathbb{E}[U_i^2|\mathscr{F}_{i-1}]$ a.s. In addition

$$\mathbb{E}[U_i^2|\mathscr{F}_{i-1}] = \mathbb{E}\left[\left(a(i,\pi(i)) - \frac{1}{n-i}\sum_{j=i+1}^{p} a(j,\pi(i))\right)^2\bigg|\mathscr{F}_{i-1}\right],$$

$$= \frac{1}{n-i+1}\sum_{k\in I_{i-1}}\left(a(i,k) - \frac{1}{n-i}\sum_{j=i+1}^{p} a(j,k)\right)^2,$$

$$\leqslant \frac{2}{n-i+1}\sum_{k\in I_{i-1}} a^2(i,k) + \left(\frac{1}{n-i}\sum_{j=i+1}^{p} a(j,k)\right)^2,$$

which implies that

$$\mathbb{E}[\Delta M_i^2|\mathscr{F}_{i-1}] \leqslant 2V_i + 2W_i \qquad \text{a.s.} \qquad (4.15)$$

where

$$V_i = m_{i-1}(a^2(i,*)), \qquad W_i = m_{i-1}\left(\left(\frac{1}{n-i}\sum_{j=i+1}^{p} a(j,*)\right)^2\right).$$

It now remains to find sharp upper bounds for the sums of (V_i) and (W_i). On the one hand, Chebyshev's sum inequality states that if (x_n) and (y_n) are nondecreasing and nonincreasing sequences, then

$$\frac{1}{n}\sum_{i=1}^{n} x_i y_i \leqslant \left(\frac{1}{n}\sum_{i=1}^{n} x_i\right)\left(\frac{1}{n}\sum_{i=1}^{n} y_i\right).$$

Consequently, as the sequence (A_i) is nonincreasing and, for any $1 \leqslant i \leqslant p$,

$$V_i = \frac{1}{n-i+1}\sum_{k\in I_{i-1}} a^2(i,k) \leqslant \frac{A_i}{n-i+1},$$

we obtain from Chebyshev's sum inequality that

$$\sum_{i=1}^{p} V_i \leqslant \sum_{i=1}^{p} \frac{A_i}{n-i+1} \leqslant \frac{\mathscr{B}}{p}\sum_{i=1}^{p} \frac{1}{n-i+1},$$

$$\leqslant \frac{\mathscr{B}}{p}\int_{1}^{p+1} \frac{dx}{n-x+1} = \frac{\mathscr{B}}{p}\log\left(\frac{n}{n-p}\right). \qquad (4.16)$$

On the other hand, by Cauchy-Schwarz inequality, we have for all $1 \leqslant i \leqslant p-1$,

$$
\begin{aligned}
W_i &= \frac{(p-i)^2}{(n-i)^2} m_{i-1}\left(\left(\frac{1}{p-i}\sum_{j=i+1}^{p} a(j,*)\right)^2\right), \\
&\leqslant \frac{(p-i)^2}{(n-i)^2} m_{i-1}\left(\frac{1}{p-i}\sum_{j=i+1}^{p} a^2(j,*)\right), \\
&\leqslant \frac{1}{n-i+1}\frac{(p-i)}{(n-i)^2}\sum_{j=i+1}^{p} A_j, \\
&\leqslant \frac{p}{n}\frac{1}{(n-i+1)(n-i)}\sum_{j=i+1}^{p} A_j, \\
&= \frac{p}{n}\left(\frac{1}{n-i}-\frac{1}{n-i+1}\right)\sum_{j=i+1}^{p} A_j.
\end{aligned}
$$

Summing over i, we obtain from (4.16) that

$$
\begin{aligned}
\sum_{i=1}^{p-1} W_i &\leqslant \frac{p}{n}\sum_{i=1}^{p-1}\sum_{j=i+1}^{p}\left(\frac{1}{n-i}-\frac{1}{n-i+1}\right)A_j, \\
&= \frac{p}{n}\sum_{j=2}^{p}\left(\frac{1}{n-j+1}-\frac{1}{n}\right)A_j, \\
&\leqslant \frac{p}{n}\sum_{j=1}^{p}\frac{A_j}{n-j+1}-\frac{p\mathscr{B}}{n^2}, \\
&\leqslant \frac{\mathscr{B}}{n}\log\left(\frac{n}{n-p}\right)-\frac{p\mathscr{B}}{n^2}. \tag{4.17}
\end{aligned}
$$

Therefore, putting together (4.15), (4.16), and (4.17), we obtain that

$$
\langle M\rangle_p \leqslant -2\mathscr{B}\left(\frac{1}{p}+\frac{1}{n}\right)\log\left(\frac{n-p}{n}\right)-2\mathscr{B}\frac{p}{n^2}=\frac{2\theta\mathscr{B}}{n} \qquad \text{a.s.} \tag{4.18}
$$

where

$$
\alpha = \frac{p}{n} \qquad \text{and} \qquad \theta = -\alpha - \frac{1+\alpha}{\alpha}\log(1-\alpha).
$$

Finally, (4.10) clearly follows from (4.11), (4.12), and (4.18), which completes the proof of Lemma 4.6. $\qquad\square$

Proof of Theorem 4.3. Denote by p some integer between 1 and $n-1$. We split S_n into two terms, $S_n = P_n + Q_n$ where

$$
P_n = \sum_{i=1}^{p} a_n(i,\pi_n(i)) \qquad \text{and} \qquad Q_n = \sum_{i=p+1}^{n} a_n(i,\pi_n(i)).
$$

It follows from Lemma 4.6 that for any positive x,

$$\mathbb{P}(|S_n - \mathbb{E}[S_n]| \geq x) \leq \mathbb{P}(|P_n - \mathbb{E}[P_n]| \geq x/2) + \mathbb{P}(|Q_n - \mathbb{E}[Q_n]| \geq x/2),$$

$$\leq 2\exp\left(-\frac{x^2}{16(\zeta v_n + x m_a/3)}\right) + 2\exp\left(-\frac{x^2}{16(\xi v_n + x m_a/3)}\right)$$

where v_n is given by (4.9), $\zeta = \varphi(\alpha)$, $\xi = \varphi(1-\alpha)$ with $\alpha = p/n$ and φ is the function defined, for all t in $]0,1[$, by

$$\varphi(t) = -t - \frac{1+t}{t}\log(1-t).$$

It is not hard to see that ζ is an increasing function of α, while ξ is a decreasing function of α. Moreover,

$$\zeta = \xi \iff \varphi(\alpha) = \varphi(1-\alpha) \iff \alpha = \frac{1}{2}.$$

It means that p should be chosen as the integer part of $n/2$. If n is even, then $\alpha = 1/2$ and $\zeta = \xi = 3\log(2) - 1/2$. If n is odd, then

$$\alpha = \frac{1}{2} - \frac{1}{2n} \qquad \text{and} \qquad \zeta \leq \xi \leq \theta$$

where $\theta = \varphi(2/3)$. In both cases, we obtain that for any positive x,

$$\mathbb{P}(|S_n - \mathbb{E}[S_n]| \geq x) \leq 4\exp\left(-\frac{x^2}{16(\theta v_n + x m_a/3)}\right)$$

which achieves the proof of Theorem 4.3. $\qquad\qquad\qquad\qquad\qquad\square$

4.3 Empirical periodogram

Let (X_n) be a centered stationary real Gaussian process. We assume that the process (X_n) has a positive continuous spectral density g bounded on the torus $\mathbb{T} = [-\pi, \pi[$, which means that

$$\mathbb{E}[X_j X_k] = \frac{1}{2\pi}\int_{\mathbb{T}} \exp(i(j-k)x)g(x)\,dx.$$

Our goal is to provide concentration inequalities for the integrated periodogram defined, for any bounded continuous real function f on the torus $\mathbb{T}$, by

$$\mathcal{W}_n(f) = \frac{1}{2\pi n}\int_{\mathbb{T}} f(x)\left|\sum_{j=1}^{n} X_j \exp(ijx)\right|^2 dx. \tag{4.19}$$

It is well known [6, 8] that

$$\lim_{n\to\infty} \mathscr{W}_n(f) = \frac{1}{2\pi} \int_{\mathbb{T}} f(x)g(x)\,dx \qquad \text{a.s.}$$

In addition, we also have

$$\sqrt{n}\Big(\mathscr{W}_n(f) - m(f)\Big) \xrightarrow{\mathcal{L}} \mathcal{N}(0, \sigma^2(f))$$

where

$$m(f) = \frac{1}{2\pi} \int_{\mathbb{T}} f(x)g(x)\,dx \quad \text{and} \quad \sigma^2(f) = \frac{1}{\pi} \int_{\mathbb{T}} f^2(x)g^2(x)\,dx.$$

The covariance matrix associated with the random vector $X^{(n)} = (X_1,\ldots,X_n)^t$ is given by $\mathbb{E}[X^{(n)}X^{(n)t}] = T_n(g)$ where $T_n(g)$ stands for the Toeplitz matrix of order n associated with g, that is the square matrix of order n composed of the Fourier coefficients of g. Consequently, it immediately follows from (4.19) that

$$\mathscr{W}_n(f) = \frac{1}{n}X^{(n)t}T_n(f)X^{(n)} = \frac{1}{n}Y^{(n)t}T_n(g)^{1/2}T_n(f)T_n(g)^{1/2}Y^{(n)} \qquad (4.20)$$

where the vector $Y^{(n)}$ has a Gaussian $\mathcal{N}(0, I_n)$ distribution. Hence, an orthogonal change of basis leads to the decomposition

$$\mathscr{W}_n(f) = \frac{1}{n} \sum_{k=1}^{n} \lambda_k^n Z_k^n \qquad (4.21)$$

where $Z_1^n,\ldots,Z_n^n$ is a sequence of independent random variables with the same $\chi^2(1)$ distribution, and $\lambda_1^n \leqslant \ldots \leqslant \lambda_n^n$ are the eigenvalues of $T_n(f)T_n(g)$ arranged by increasing order. It follows from an extension [3] of the classical Szegö theorem that the empirical spectral measure of the product $T_n(f)T_n(g)$ converges to P_{fg} which is the image probability of the uniform measure on the torus $\mathbb{T}$ by the product fg. In other words, for any bounded continuous real function h,

$$\lim_{n\to\infty} \frac{1}{n} \sum_{k=1}^{n} h(\lambda_k^n) = \frac{1}{2\pi} \int_{\mathbb{T}} h(f(x)g(x))\,dx. \qquad (4.22)$$

In addition, it was recently shown in [5] that λ_n^n and λ_1^n both converge to the extrema of the spectrum of the product $T(f)T(g)$

$$\lambda_{\max}(f,g) = \max \sigma\big(T(f)T(g)\big),$$
$$\lambda_{\min}(f,g) = \min \sigma\big(T(f)T(g)\big).$$

One can observe that $T_n(f)$ and $T_n(g)$ are the finite sections of order $n \geqslant 1$ of the Toeplitz operators $T(f)$ and $T(g)$ and that $T(f)T(g)$ is of course different from the

Toeplitz operator of the product fg, namely $T(fg)$. Let L_{fg} be the limiting cumulant generating function of $\mathcal{W}_n(f)$,

$$L_{fg}(t) = -\frac{1}{4\pi} \int_{\mathbb{T}} \log(1 - 2tf(x)g(x))\, dx.$$

Denote by I_{fg} its Legendre-Fenchel transform

$$I_{fg}(x) = \sup_{t \in \mathbb{R}} \{xt - L_{fg}(t)\}.$$

The large deviation principle for the sequence $(\mathcal{W}_n(f))$ was established in [3] and recently improved in [5]. More precisely, the sequence $(\mathcal{W}_n(f))$ satisfies a large deviation principle with good rate function J_{fg} given, for all $x \in \mathbb{R}$, by

$$J_{fg}(x) = \begin{cases} I_{fg}(a) + \dfrac{1}{2\lambda_{\min}(f,g)}(x-a) & \text{if } x \in\,]-\infty, a] \\[2ex] I_{fg}(x) & \text{if } x \in\,]a, b[\\[2ex] I_{fg}(b) + \dfrac{1}{2\lambda_{\max}(f,g)}(x-b) & \text{if } x \in [b, +\infty[\end{cases} \qquad (4.23)$$

where a and b are the extended real numbers given by

$$a = L'_{fg}\left(\frac{1}{2\lambda_{\min}(f,g)}\right)$$

if $\lambda_{\min}(f,g) < 0$ and $\lambda_{\min}(f,g) < \inf(fg)$, $a = -\infty$ otherwise, while

$$b = L'_{fg}\left(\frac{1}{2\lambda_{\max}(f,g)}\right)$$

if $\lambda_{\max}(f,g) > 0$ and $\lambda_{\max}(f,g) > \sup(fg)$, $b = +\infty$ otherwise. We shall now provide a concentration inequality for $\mathcal{W}_n(f)$. For the sake of simplicity, we now assume that f is a positive function. For any positive x, denote

$$\Phi_R(x) = x - \log(1+x)$$

while, for any $x \in\,]0,1[$, let

$$\Phi_L(x) = -x - \log(1-x).$$

Theorem 4.7. *For any positive x,*

$$\mathbb{P}\left(\mathcal{W}_n(f) - \mathbb{E}[\mathcal{W}_n(f)] \geq x\right) \leq \exp\left(-\frac{n^2 Var(\mathcal{W}_n(f))}{(2\lambda_n^n)^2} \Phi_R\left(\frac{2\lambda_n^n x}{n Var(\mathcal{W}_n(f))}\right)\right). \qquad (4.24)$$

In addition, for any x in $]0, 1[$,

$$\mathbb{P}\Big(\mathscr{W}_n(f) - \mathbb{E}[\mathscr{W}_n(f)] \leqslant -x\mathbb{E}[\mathscr{W}_n(f)]\Big) \leqslant \exp\Big(-\frac{\big(\mathbb{E}[\mathscr{W}_n(f)]\big)^2}{Var(\mathscr{W}_n(f))}\Phi_L(x)\Big). \qquad (4.25)$$

Remark 4.8. We already saw that, for any positive x,

$$\Phi_R(x) = x - \log(1+x) \geqslant \frac{x^2}{1+x+\sqrt{1+2x}}$$

while, for any $x \in]0, 1[$,

$$\Phi_L(x) = -x - \log(1-x) \geqslant \frac{x^2}{2}.$$

Consequently, inequalities (4.24) and (4.25) improve the exponential inequalities for weighted sum of chi-square distributions given in Lemma 1 of [14]. Moreover, for any positive x,

$$\mathbb{P}\Big(\mathscr{W}_n(f) - \mathbb{E}[\mathscr{W}_n(f)] \geqslant 2\lambda_n^n x + \sqrt{2nx\mathrm{Var}(\mathscr{W}_n(f))}\Big) \leqslant \exp(-nx)$$

and

$$\mathbb{P}\Big(\mathscr{W}_n(f) - \mathbb{E}[\mathscr{W}_n(f)] \leqslant -\sqrt{2nx\mathrm{Var}(\mathscr{W}_n(f))}\Big) \leqslant \exp(-nx).$$

Proof. The proof of Theorem 4.7 mainly relies on Theorem 2.57. As a matter of fact, we obtain from (4.21) that

$$\mathscr{W}_n(f) = \frac{1}{n}S_n(f) \qquad \text{where} \qquad S_n(f) = \sum_{k=1}^{n} \lambda_k^n Z_k^n.$$

Hence,

$$S_n(f) = \sum_{k=1}^{n} \xi_k^n$$

where ξ_k^n has the $\Gamma(a_k, b_k)$ distribution with $a_k = 1/2$ and $b_k = 2\lambda_k^n$. Consequently, (4.24) and (4.25) immediately follow from (2.103) and (2.105), respectively. $\qquad\square$

4.4 Random matrices

Let $X_1, \ldots, X_n$ be a finite sequence of independent centered random vectors of $\mathbb{R}^p$ given by

$$X_1 = \begin{pmatrix} X_{11} \\ X_{21} \\ \vdots \\ X_{p1} \end{pmatrix}, \quad X_2 = \begin{pmatrix} X_{12} \\ X_{22} \\ \vdots \\ X_{p2} \end{pmatrix}, \quad \dots \quad X_n = \begin{pmatrix} X_{1n} \\ X_{2n} \\ \vdots \\ X_{pn} \end{pmatrix}.$$

The empirical covariance matrix associated with $X_1, \dots, X_n$ is the Wishart random matrix given by

$$\mathbf{A}_p = \frac{1}{n}\left(X_1 X_1^t + \cdots + X_n X_n^t\right) = \frac{1}{n} X X^t$$

where X is the rectangular $p \times n$ matrix

$$X = \begin{pmatrix} X_{11} & X_{12} & \cdots & \cdots & X_{1n} \\ X_{21} & X_{22} & \cdots & \cdots & X_{2n} \\ \vdots & \vdots & & & \vdots \\ X_{p1} & X_{p2} & \cdots & \cdots & X_{pn} \end{pmatrix}.$$

We refer the reader to the books [1] and [2] for more insight on the theory of random matrices. Our objective here is to establish concentration inequalities for symmetric functionals of the eigenvalues

$$Z = f(\lambda_1, \dots, \lambda_p)$$

where $\lambda_1, \dots, \lambda_p$ are the eigenvalues of the random matrix $\mathbf{A}_p$. The amazing point is that these inequalities are essentially a consequence of mechanical application of Theorem 2.62.

In this context, the dependence is usually taken into account in two different ways: one assumes either that the random vectors $X_1, \dots, X_n$ are independent and identically distributed but each vector has dependent coordinates or that the random matrix X can be written, for some real mixture square matrix Γ of order p,

$$X = \Gamma \xi$$

where ξ is a random rectangular $p \times n$ matrix with independent entries, which means that for all $1 \leqslant k \leqslant n$, X_k has the specific form $\Gamma \xi_k$. This second way is thus more restrictive, but is natural from a theoretical point of view since one obtains essentially the same results as in the case $\Gamma = \mathbf{I}_p$, see Chapter 8 of [2]. We shall consider both cases, the first one being much easier, but leading to much larger bounds when p is large.

We refer the reader to Delyon [10] for more details on the results below. Our strategy mainly relies on Theorem 4.9 which is a direct consequence of Theorem 2.62.

Theorem 4.9. *Let $X_1, \dots, X_n$ be a finite sequence of zero-mean independent random variables with support contained in some measured space E. Let F be a measurable function defined on E^n with real values. For all $1 \leqslant k \leqslant n$, and for $a \in E$, denote*

$$X^{(k)}(a) = (X_1, \dots, X_{k-1}, a, X_{k+1}, \dots, X_n)$$

and

$$C_k = \sup_{a,b \in E} \left(F(X^{(k)}(b)) - F(X^{(k)}(a)) \right).$$

Then,

$$\mathbb{E}[\exp(F(X) - \mathbb{E}[F(X)])] \leqslant \exp\left(\frac{\mathscr{C}_n}{8}\right) \quad \text{where} \quad \mathscr{C}_n = \left\| \sum_{k=1}^{n} C_k^2 \right\|_{\infty}. \quad (4.26)$$

Remark 4.10. The key point in (4.26) is that the infinity norm is outside the sum.

4.4.1 Independent entries with mixture matrix

We focus our attention on the situation where

$$\mathbf{A}_p = \frac{1}{n} \Gamma \xi \xi^t \Gamma^t$$

where the mixture matrix Γ is a deterministic real square matrix of order p and ξ is a random rectangular $p \times n$ matrix with independent entries. Denote

$$\delta = \sup_{ij} \left(\text{esssup} \, \xi_{ij} - \text{essinf} \, \xi_{ij} \right) \quad \text{and} \quad \sigma = \|\Gamma\|^2 \sup_{ij} \|\xi_{ij}\|_{\infty}.$$

According to Theorem 9.10 of [2], when f is an additive function of the eigenvalues, the asymptotic variance of Z as n tends to infinity and p/n converges to some limit $\gamma > 0$, exists as a rather complicated function of γ. Hence, a reasonable objective would be to obtain concentration inequalities like

$$\mathbb{P}(|Z - \mathbb{E}[Z]| \geqslant x) \leqslant \exp\left(-h(p/n)x^2\right) \quad (4.27)$$

for some function h. The next theorem provides such a bound.

Theorem 4.11. *Let $Z = f(\lambda_1, \ldots, \lambda_p)$ where $\lambda_1, \ldots, \lambda_p$ are the eigenvalues of the random matrix $\mathbf{A}_p$ and f is a twice differentiable symmetric function defined on $\mathbb{R}^p$ with real values. Then,*

$$\mathbb{E}[\exp(Z - \mathbb{E}[Z])] \leqslant \exp\left(\frac{p}{2n} \sigma^2 \delta^2 \left(3\alpha + \frac{2\beta p}{n} \sigma \delta \right)^2 \right) \quad (4.28)$$

where α is the infinity norm of the gradient of f and β is the matrix norm of the Hessian matrix of f. Consequently, for any positive x,

$$\mathbb{P}(|Z - \mathbb{E}[Z]| \geqslant x) \leqslant 2 \exp\left(-\frac{n}{2p} \frac{x^2}{\sigma^2 \delta^2} \left(3\alpha + \frac{2p}{n} \sigma \delta \beta \right)^{-2} \right). \quad (4.29)$$

Remark 4.12. In the particular case where $\Gamma = \mathrm{I}_p$ and f is an additive function of the eigenvalues

$$f(\lambda_1,\ldots,\lambda_p) = \sum_{k=1}^{p} \varphi(\lambda_k),$$

Guionnet and Zeitouni [11] have shown that, for any positive x,

$$\mathbb{P}(|Z - \mathbb{E}[Z]| \geqslant x) \leqslant 4\exp\left(-\frac{x^2}{g^2}\right) \qquad (4.30)$$

for some reasonable constant g with order of magnitude

$$g \sim \sigma\|x\varphi'(x^2)\|_\infty\left(1+\frac{p}{n}\right)$$

and under the assumption of convexity of the function $x \mapsto \varphi(x^2)$. A similar result is also given without this convexity assumption but a logarithmic Sobolev inequality is required for the densities of the ξ_{ij}s. We deduce from Theorem 4.11 that, in this particular case,

$$\mathbb{P}(|Z - \mathbb{E}[Z]| \geqslant x) \leqslant 2\exp\left(-\frac{n}{2p}\frac{x^2}{\zeta^2}\right) \qquad (4.31)$$

where

$$\zeta = \sigma\delta\left(\|\varphi'\|_\infty + \frac{p}{n}\sigma\delta\|\varphi''\|_\infty\right).$$

In the case where n tends to infinity and $p = \gamma n$, (4.30) has the advantage, over (4.31), of not involving second derivatives of the function φ. However, when $p \ll n$, inequality (4.31) is always sharper as soon as φ'' is bounded.

4.4.2 Independent columns with dependent coordinates

We now deal with the situation where

$$\mathbf{A}_p = \frac{1}{n}XX^t$$

and the random column vectors $X_1,\ldots,X_n$ of X are independent and identically distributed but each vector has dependent coordinates. Let $E \subset \mathbb{R}^p$ be a set containing the support of the distribution of each column vectors. Denote

$$\delta = \sup_{a,b\in E}\|a-b\| \quad \text{and} \quad \sigma = \sup_{a\in E}\|a\|.$$

Theorem 4.13. *Let* $Z = f(\lambda_1,\ldots,\lambda_p)$ *where* $\lambda_1,\ldots,\lambda_p$ *are the eigenvalues of the random matrix* $\mathbf{A}_p$ *and* f *is a differentiable symmetric function defined on* $\mathbb{R}^p$ *with real values. Then,*

$$\mathbb{E}[\exp(Z - \mathbb{E}[Z])] \leqslant \exp\left(\frac{9\sigma^2\delta^2\alpha^2}{2n}\right) \tag{4.32}$$

where α is the infinity norm of the gradient of f. Consequently, for any positive x,

$$\mathbb{P}(|Z - \mathbb{E}[Z]| \geqslant x) \leqslant 2\exp\left(-\frac{nx^2}{18\sigma^2\delta^2\alpha^2}\right). \tag{4.33}$$

Remark 4.14. One can observe that σ and δ are expected to have order $\sqrt{p}$. In the particular case where $|X_{ij}| \leqslant 1$ for all i, j and f is an additive function of the eigenvalues

$$f(\lambda_1, \ldots, \lambda_p) = \sum_{k=1}^{p} \varphi(\lambda_k),$$

it was proved by Guntuboyina and Leeb [12] that, for any positive x,

$$\mathbb{P}(|Z - \mathbb{E}[Z]| \geqslant x) \leqslant 2\exp\left(-\frac{2x^2}{nV^2}\right) \tag{4.34}$$

where V is the total variation of the function φ on $\mathbb{R}$. This bound is essentially optimal in the context of independent columns if p and n have the same order of magnitude, see Example 3 in [12]. For comparison, it follows from Theorem 4.13 with $\sigma^2 = p$ and $\delta^2 = 4p$ that, for any positive x,

$$\mathbb{P}(|Z - \mathbb{E}[Z] \geqslant x) \leqslant \exp\left(-\frac{nx^2}{72p^2c}\right)$$

where $c = \|\varphi'\|_\infty^2$. In the independent case, this improves on Corollary 1.8 of [11] only if p^2 is significantly smaller than n. It also improves on (4.34) if $p \ll n$ since in this case $p^2/n \ll n$.

4.4.3 Proofs

In order to prove Theorem 4.11 and Theorem 4.13, we make use of the following lemma, which is established in the Appendix of [10].

Lemma 4.15. *Let I be an open interval of $\mathbb{R}$. For all $t \in I$, denote $\Phi(t) = M(t)$ where $M(t)$ is a parameterized symmetric square matrix of order p and assume that Φ belongs to C^{3p}. Then, it exists a twice differentiable parametrization of the eigenvalues $\Lambda(t) = (\lambda_1(t), \ldots \lambda_p(t))$ of the matrices $M(t)$. Let f be a twice differentiable symmetric function defined on some cube $Q = [a,b]^p$ containing the eigenvalues of the matrices $M(t), t \in I$. Then, the function*

$$\varphi(t) = f(\Lambda(t))$$

is twice differentiable on I. In addition, for any $t \in I$,

$$\dot{\varphi}(t) = \sum_{k=1}^{p} d_k^{-1}(t) \frac{\partial f}{\partial \lambda_k}(\Lambda(t)) Tr(\dot{M}(t) P_{\lambda_k(t)}), \tag{4.35}$$

$$|\ddot{\varphi}(t)| \leqslant \gamma Tr(\dot{M}^2(t)) + \sup_k \left| \frac{\partial g}{\partial \lambda_k}(\Lambda(t)) \right| \sum_{i=1}^{p} |v_i(t)| \tag{4.36}$$

where $P_{\lambda_k(t)}$ is the orthogonal projection on the eigenspace $E_{\lambda_k(t)}$ of the eigenvalue $\lambda_k(t)$, $d_k(t)$ is the dimension of $E_{\lambda_k(t)}$, $v_1(t), \ldots, v_p(t)$ are the eigenvalues of $\ddot{M}(t)$, and γ is the matrix norm of the Hessian matrix of f,

$$\gamma = \sup_{t \in I} \sup_{\Lambda(t) \in Q} \|\nabla^2 f(\Lambda(t))\|.$$

Proof of Theorem 4.11. Let F be the function defined on $\mathbb{R}^{p \times n}$ by

$$F(\xi) = f(\lambda_1, \ldots, \lambda_p).$$

We have to estimate

$$\sum_{i=1}^{p} \sum_{j=1}^{n} \sup_{a,b} (F(\xi^{(ij)}(b)) - F(\xi^{(ij)}(a))^2$$

with the notations of Theorem 4.9. If we set

$$\delta = \sup_{ij} \left(\operatorname{esssup} \xi_{ij} - \operatorname{essinf} \xi_{ij} \right),$$

the Taylor formula implies that

$$\left| F(\xi^{(ij)}(b)) - F(\xi^{(ij)}(a)) \right| \leqslant \delta \left| \frac{\partial F}{\partial \xi_{ij}}(\xi) \right| + \delta^2 \sup_{k,l,\zeta} \left\| \frac{\partial^2 F}{\partial \xi_{kl}^2}(\zeta) \right\|_\infty = A_{ij} + B.$$

We shall estimate these quantities with the help of Lemma 4.15 where the variable t coincides with ξ_{ij}. We have for any matrix P,

$$\operatorname{Tr}\left(P \frac{\partial \mathbf{A}_p}{\partial \xi_{ij}} \right) = \frac{1}{n} \frac{\partial}{\partial \xi_{ij}} Tr(\xi^t \Gamma^t P \Gamma \xi) = \frac{1}{n} (\Gamma^t (P + P^t) \Gamma \xi)_{ij}. \tag{4.37}$$

Hence, equation (4.35) can be rewritten as

$$\frac{\partial F(\xi)}{\partial \xi_{ij}} = \frac{2}{n} \sum_{k=1}^{p} d_k^{-1} \frac{\partial f(\lambda)}{\partial \lambda_k} (\Gamma^t P_{\lambda_k} \Gamma \xi)_{ij}$$

where P_{λ_k} is the orthogonal projection on the eigenspace of $\mathbf{A}_p$ corresponding to λ_k, and d_k is the dimension of this space. It happens that we can get a good bound for a partial Euclidean norm of the gradient of F. Denote

$$f_k = \frac{\partial f(\lambda)}{\partial \lambda_k} \quad \text{and} \quad V_k = P_{\lambda_k}\Gamma\xi_j$$

where ξ_j stands for the jth column vector of the random rectangular $p \times n$ matrix ξ. Notice that in case of eigenvalues of higher multiplicities, some V_ks are repeated, but if $V_k = V_\ell$, $\lambda_k = \lambda_\ell$, one has also $f_k = f_\ell$ by symmetry of f. Consequently, denoting by K a set of indices such that $\sum_{k \in K} P_{\lambda_k} = I_p$, we have by orthogonality of $V_1, \ldots, V_p$,

$$\frac{n^2}{4}\sum_{i=1}^{p}\left(\frac{\partial F(\xi)}{\partial \xi_{ij}}\right)^2 = \sum_{i=1}^{p}\left(\sum_{k=1}^{p} f_k d_k^{-1}(\Gamma^t P_{\lambda_k}\Gamma\xi)_{ij}\right)^2 = \sum_{i=1}^{p}\left(\sum_{k=1}^{p} f_k d_k^{-1}(\Gamma^t V_k)_i\right)^2$$

$$= \sum_{i=1}^{p}\left(\sum_{k \in K} f_k(\Gamma^t V_k)_i\right)^2 = \left\|\Gamma^t\left(\sum_{k \in K} f_k V_k\right)\right\|^2$$

$$\leqslant \|\Gamma\|^2\left\|\sum_{k \in K} f_k V_k\right\|^2 = \|\Gamma\|^2 \sum_{k \in K} f_k^2 \|V_k\|^2$$

$$\leqslant \|\Gamma\|^2\alpha^2 \sum_{k \in K} \|V_k\|^2 = \alpha^2\|\Gamma\|^2\Big\|\sum_{k \in K} V_k\Big\|^2$$

$$= \alpha^2\|\Gamma\|^2\|\Gamma\xi_j\|^2 \leqslant \alpha^2\|\Gamma\|^4\|\xi_j\|^2 \leqslant \alpha^2\sigma^2 p.$$

For the estimation of the second derivatives, equation (4.36) can be rewritten as

$$\left|\frac{\partial^2 F(\xi)}{\partial \xi_{ij}^2}\right| \leqslant \alpha \sum_{k=1}^{p} |v_k| + \beta\,\mathrm{Tr}(\mathbf{B}_p^2)$$

where $v_1, \ldots, v_p$ are the eigenvalues of the matrix

$$\frac{\partial^2 \mathbf{A}_p}{\partial \xi_{ij}^2} \quad \text{and} \quad \mathbf{B}_p = \frac{\partial \mathbf{A}_p}{\partial \xi_{ij}}.$$

On the one hand, we clearly have

$$\frac{\partial^2 \mathbf{A}_p}{\partial \xi_{ij}^2} = \frac{2}{n}\Gamma_i\Gamma_i^t.$$

It means that $v_1, \ldots, v_p$ are all zero except for one, and $2n^{-1}\|\Gamma_i\|^2 \leqslant 2n^{-1}\|\Gamma\|^2$. On the other hand, it follows from (4.37) with $P = \mathbf{B}_p = P^t$ that

$$\mathrm{Tr}(\mathbf{B}_p^2) \leqslant \frac{2}{n}\|\Gamma\|^2(p\sup_{ij}\|\xi_{ij}\|^2)^{1/2}\mathrm{Tr}(\mathbf{B}_p^2)^{1/2}$$

which leads to $\mathrm{Tr}(\mathbf{B}_p^2) \leqslant 4p\sigma^2/n^2$. Finally, we obtain that

$$\left|\frac{\partial^2 F(\xi)}{\partial \xi_{ij}^2}\right| \leqslant \frac{2\alpha\|\Gamma\|^2}{n} + \frac{4\beta p\sigma^2}{n^2}.$$

Consequently, we deduce from the above calculation that, for any $1 \leqslant j \leqslant n$,

$$\sum_{i=1}^{p}\left(F(\xi^{(ij)}(b)) - F(\xi^{(ij)}(a))\right)^2 \leqslant \sum_{i=1}^{p} A_{ij}^2 + 2B\sum_{i=1}^{p} A_{ij} + pB^2$$

$$\leqslant \sum_{i=1}^{p} A_{ij}^2 + 2B\sqrt{p}\left(\sum_{i=1}^{p} A_{ij}^2\right)^{1/2} + pB^2 = \left(\left(\sum_{i=1}^{p} A_{ij}^2\right)^{1/2} + B\sqrt{p}\right)^2$$

$$\leqslant \left(\frac{2\alpha\delta\sigma\sqrt{p}}{n} + \frac{\delta^2}{n}\left(\frac{4\beta p\sigma^2}{n} + 2\alpha\|\Gamma\|^2\right)\sqrt{p}\right)^2$$

$$\leqslant \left(\frac{6\alpha\delta\sigma\sqrt{p}}{n} + \frac{4\beta p\sigma^2\delta^2}{n^2}\sqrt{p}\right)^2 = \frac{4p\delta^2\sigma^2}{n^2}\left(3\alpha + \frac{2\beta p}{n}\delta\sigma\right)^2 \quad (4.38)$$

as $\delta\|\Gamma\|^2 \leqslant 2\sigma$. Therefore, inequality (4.28) clearly follows from (4.26) and (4.38). Finally, we easily deduce (4.29) from (4.28) via the usual Chernoff calculation, which completes the proof of Theorem 4.11. $\qquad\qquad\square$

Proof of Theorem 4.13. Let F be the function defined on $\mathbb{R}^{p\times n}$ by

$$F(X) = f(\lambda_1,\ldots,\lambda_p).$$

We consider the rectangular matrix X as a family of random column vectors $X_1,\ldots,X_n$ and we use the notations of Theorem 4.9. As before, it is necessary to estimate

$$\sum_{j=1}^{n}\sup_{a,b}(F(X^{(j)}(b)) - F(X^{(j)}(a))^2$$

where a and b are deterministic vectors of $\mathbb{R}^p$. We have from the Taylor formula

$$\left|F(X^{(j)}(b)) - F(X^{(j)}(a))\right| \leqslant \sup_{0\leqslant t\leqslant 1}\left|\frac{\partial F(X + t\Delta(a,b))}{\partial t}\right|$$

where $\Delta(a,b) = X^{(j)}(b) - X^{(j)}(a)$. The matrix $\Delta(a,b)$ vanishes except on its j-th column which is the vector $c = b - a$. We shall estimate upper bound with the help of Lemma 4.15 where

$$M(t) = \frac{1}{n}(X + t\Delta(a,b))(X + t\Delta(a,b))^t.$$

We have, for all $t \in [0,1]$,

$$\dot{M}(t) = \frac{1}{n}\left(X\Delta(a,b)^t + \Delta(a,b)X^t + 2t\Delta(a,b)\Delta(a,b)^t\right) = \frac{1}{n}\left(cX_j^t + X_jc^t + 2tcc^t\right).$$

It follows from (4.35) that, for any $t \in [0,1]$,

$$\frac{\partial F(X + t\Delta(a,b))}{\partial t} = \frac{2}{n}\sum_{k=1}^{p} d_k^{-1}\frac{\partial f(\lambda)}{\partial \lambda_k}\langle P_{\lambda_k}c, X_j + tc\rangle$$

where $\lambda_1(t),\ldots,\lambda_p(t)$ are the eigenvalues of the matrices $M(t)$. Consequently, as P_{λ_k} is an orthogonal projector, we deduce from Cauchy-Schwarz's inequality that

$$\begin{aligned}
\left|\frac{\partial F(X + t\Delta(a,b))}{\partial t}\right| &\leqslant \frac{2\alpha}{n}\sum_{k\in K}\|P_{\lambda_k}c\|\|P_{\lambda_k}(X_j + tc)\| \\
&\leqslant \frac{2\alpha}{n}\left(\sum_{k\in K}\|P_{\lambda_k}c\|^2\right)^{1/2}\left(\sum_{k\in K}\|P_{\lambda_k}(X_j + tc)\|^2\right)^{1/2} \\
&\leqslant \frac{2\alpha}{n}\|c\|\|X_j + tc\| \leqslant \frac{6\alpha}{n}\delta\sigma.
\end{aligned}$$

Therefore, we obtain that, for all $1 \leqslant j \leqslant n$

$$\left|F(X^{(j)}(b)) - F(X^{(j)}(a))\right|^2 \leqslant \frac{16\alpha^2 p^2}{n^2}\delta^2\sigma^2$$

which immediately leads to (4.32). Finally, (4.33) follows from (4.32) via the usual Chernoff calculation, which achieves the proof of Theorem 4.13. $\qquad\square$

References

1. Anderson, G. W., Guionnet A. and Zeitouni, O.: An introduction to random matrices. Cambridge University Press, New York, (2010)
2. Bai, Z. and Silverstein, J. W.: Spectral analysis of large dimensional random matrices. Springer Series in Statistics. Springer, New York, second edition, (2009)
3. Bercu, B., Gamboa, F. and Rouault, A.: Large deviations for quadratic forms of stationary Gaussian processes. Stochastic Process. Appl. **71** 75–90 (1997)
4. Bercu, B. and Touati, A.: Exponential inequalities for self-normalized martingales with applications. Ann. Appl. Probab. **18** 1848–1869 (2008)
5. Bercu, B., Bony, J. F. and Bruneau, V.: Large deviations for Gaussian stationary processes and semi-classical analysis, Séminaire de Probabilités **44**, Lecture Notes in Mathematics 2046, 409–428 (2012).
6. Brockwell, P. J. and Davis, R. A.: Time series: Theory and methods. Springer Series in Statistics. Springer, New York, second edition, (2009)
7. Chatterjee, S.: Stein's method for concentration inequalities. Probab. Theory Related Fields **138** 305–321 (2007)
8. Dacunha-Castelle, D. and Duflo, M.: Probability and Statistics. Volume II. Springer-Verlag, New York, (1986)

9. Del Moral, P. and Rio, E.: Concentration inequalities for mean field particle models. Ann. Appl. Probab. **21**, 1017–1052 (2011)
10. Delyon, B.: Concentration inequalities for the spectral measure of random matrices. Electron. Commun. Probab. **15**, 549–561 (2010)
11. Guionnet, A. and Zeitouni, O.: Concentration of the spectral measure for large matrices. Electron. Commun. Probab. **5**, 119–136 (2000)
12. Guntuboyina, A. and Leeb, H.: Concentration of the spectral measure of large Wishart matrices with dependent entries. Electron. Commun. Probab. **14**, 334–342 (2009)
13. Hoeffding, W.: A Combinatorial central limit theorem. Ann. Math. Stat. **22** 558–566 (1951)
14. Laurent, B. and Massart, P.: Adaptive estimation of a quadratic functional by model selection. Ann. Stat. **28** 1302–1338 (2001)
15. Liptser, R. and Spokoiny, V.: Deviation probability bound for martingales with applications to statistical estimation. Statist. Probab. Lett. **46** 347–357 (2000)
16. Worms, J.: Moderate deviations for stable Markov chains and regression models. Electron. J. Probab. **4** 1–28 (1999)
17. Worms, J.: Large and moderate deviations upper bounds for the Gaussian autoregressive process. Statist. Probab. Lett. **51** 235–243 (2001)